Birdpedia

Birdpedia

A Brief Compendium of Avian Lore

Christopher W. Leahy

Illustrations by Abby McBride

PRINCETON UNIVERSITY PRESS
Princeton & Oxford

Published by Princeton University Press
41 William Street, Princeton, New Jersey 08540
6 Oxford Street, Woodstock, Oxfordshire OX20 1TR

press.princeton.edu

All Rights Reserved
ISBN 978-0-691-20966-1
ISBN (e-book) 978-0-691-21823-6

British Library Cataloging-in-Publication Data is available

Editorial: Robert Kirk and Abigail Johnson
Production Editorial: Mark Bellis
Text and Cover Design: Chris Ferrante
Production: Steve Sears
Publicity: Matthew Taylor and Kate Farquhar-Thomson
Copyeditor: Lucinda Treadwell

Cover, endpaper, and text illustrations by Abby McBride

This book has been composed in Plantin, Futura, and Windsor

Printed on acid-free paper. ∞

Printed in China

10 9 8 7 6 5 4 3 2 1

For James Baird—mentor, colleague, friend

Preface

Almost four decades ago, I published "an encyclopedic handbook of North American birdlife" titled *The Birdwatcher's Companion* (Hill and Wang, 1982). At 917 pages, it fit comfortably under the definition of a "tome." In 2002 Princeton University Press brought out a thoroughly revised and updated second edition that was even heftier, with a page count well over 1,000. My dual—and perhaps somewhat dueling—objectives for the *Companion* were: (1) That it be comprehensive, encompassing a full gamut of ornithological knowledge, from what "crepuscular" means and who Alexander Wilson was to how many species of woodpeckers there are in the world and how to cook a scoter; and (2) That this potentially daunting accumulation of bird lore, while striving for meticulous accuracy, could also be written in an accessible style that could be read for pleasure as well as information—even for fun.

Aside from a disparity in pure tonnage, the main difference between the *Companion* and the present modest volume is that the *Birdpedia*, while still composed of entries in alphabetical order, makes no claim to be "encyclopedic." It might be described as a "teaser" perhaps, aiming ideally for the kind of curious reader who has noticed that a large and growing percentage of the world's population has become fascinated—in some cases obsessed!—with birdwatching, or to use the sportier term, "birding." If people now spend billions of dollars annually on optical equipment, identification guides, bird feeding paraphernalia, and guided tours to Mongolia in search of exotic species, it might be worth looking into a little book to find out why so many otherwise sane people are staring into the trees or scanning smelly mudflats these days.

In the *Birdpedia*, you will find no exhaustive accounts of bird taxonomy or the avian digestive system or even descriptions of bird families. But there are general essays on Birdwatching and Identification that attempt to give the uninitiated a sense of what the fuss is all about; summaries of some of the more fascinating aspects of birdlife such as migration, brood parasitism, and vocal mimicry; as well as briefer, more whimsical entries calculated to provoke a smile or stretch credulity. The reader will still find a definition of "crepuscular" (not to mention "goatsucker"); still discover the identity of Alexander Wilson (not to mention Eleanora of Arborea); and still have access to scoter recipes. But the geographic coverage has been expanded beyond North America, and there is substantive material that did not appear in the *Companion*, such as Shakespeare's Birds and Birding While Black.

There are birds everywhere. Swimming below the ice in Antarctica. Nesting by the millions on Arctic tundra.

Dancing in the trees in the rain forests of Papua New Guinea. Gathering at oases in the Gobi Desert. Soaring over the highest peaks of the Andes and migrating above Mt. Everest. Chasing fish more than 1,750 feet down in the ocean depths. Sharing a meal with a pride of lions. Nesting on skyscrapers in New York City. And their distribution is by no means limited to geography. Birds are abundant in our art, in our poetry, in our music, in our myths and movies and medicine, in our food and fashions and fantasies. And in the fossil record millions of years before there were any human bones to ponder. It is this astonishing avian diversity—of form, of behavior, of interaction with our own species—that the *Birdpedia* means to deploy to turn a nagging curiosity into a compelling fascination and perhaps a new, more intimate relationship with the natural world.

It can be said that a love of birds manifests itself in three fundamental ways: (1) as pure pleasure—the first Baltimore Oriole of the spring, the cry of a curlew over the marsh; (2) as an ever-widening curiosity that leads to newfound knowledge, perhaps even to wisdom; and (3) as a concern for the fate of the world's birdlife—now gravely threatened by human recklessness—and a willingness to take action, however modest, to conserve it. My fondest hope for this modest volume of bird lore is that it might provide inspiration for all three.

Birdpedia

Red-billed Quelea

A bundance (How many birds?)

It should surprise no one that the question of how many individual birds are alive on this planet at any given moment has yet to be answered with any degree of certainty. True, there are a few highly conspicuous species, for example, Whooping Crane, whose breeding and wintering distributions have been fully discovered and whose total populations are so small that we know precisely how many individuals presently exist. But even estimates for scarce and well-studied species may have large error factors, either because it is difficult to distinguish individuals in populations of wide-ranging species, such as raptors, or because population numbers are extrapolated based on counts of singing males (e.g., most rare songbirds). The breeding populations of certain seabirds, for example, Laysan Albatross, Great Shearwater, Northern Gannet, and Roseate Tern, that nest locally in compact colonies—most of which are known—can be estimated with a high degree of accuracy simply by counting nest sites (though these counts do not include pre-breeding or "vacationing" birds). But when we consider even so limited a goal as calculating the number of Black-capped Chickadees in Massachusetts during a given month—not to mention the number of songbirds in India or the planet's total avian population—we begin to appreciate the difficulty of the task. To begin with, simply counting accurately the number of small wood-land birds in a 10-acre plot involves much patience and labor and leaves the counter with little confidence that absolute accuracy has been achieved. Once there are reasonably reliable counts for a range of species and habitats, one can begin to make some tentative extrapolations—but consider some of the variables involved:

— The populations of all birds fluctuate greatly in the course of the year, hitting a low before birds of the year have hatched—when only individuals that have survived the ravages of age, winter, and predation exist—and peaking at the end of the breeding season, when in most cases juvenile birds outnumber adults by at least severalfold.

— Even small-scale censuses show that some habitats support many more birds than others do—from as few as 13 birds (adults of all species) per 10 acres of woods or farmland to more than 1,000 individuals in the same amount of exceptionally rich habitat such as some tropical forests and wetlands.

— Bird populations in a given locality change drastically during the year, not only due to seasonal fluctuations effected by reproduction, but also due to bird movements. Areas of Arctic tundra that teem with breeding shorebirds and other birds are nearly (or totally) barren of birdlife for six or more months of the year. And the already rich resident avifauna of Central America is increased no-one-knows-how-many times by the flood of North American migrants that arrive on their wintering grounds there each fall.

— Population densities of different species vary greatly, making it impossible to extrapolate total bird populations from what is known about one or a few species. Bird populations increase and decrease in cycles and according to ecological variations. Northern finches and raptors experience alternating "booms" and "busts"—continually shifting patterns brought about by fluctuations in the amount of food available to the potential consumers.

SOME "GUESSTIMATES." Despite the inherent difficulty in counting birds, the question of "how many" exerts a powerful fascination, partly, no doubt, because there are obviously so many and partly because it would be very useful in judging the health of a given species (as well as of the whole environment) to know whether bird populations are declining drastically (now, alas, well documented for North America), holding their own, or even (certainly true in some cases) increasing. The following figures are taken from reliable sources, but are wisely hedged in some cases.

— The world's total avian population has been reckoned (by broad extrapolation) at around 100 billion—give or take some hundreds of millions!

— The total population of wild land birds in the contiguous United States at the beginning of the breeding season has been estimated at 5–6 billion birds, jumping to perhaps 20 billion with the addition of the young birds hatched in that breeding season).

— In a notably careful survey running the length of Finland and considering all habitats, the ornithologist Einari Merikallio came up with a total of 64 million adult breeding birds, with the two commonest species (Chaffinch and Willow Warbler) making up 10% of the whole.

— There are about 44 North American species of birds (scarce, local and/or conspicuous, colonial seabirds) whose total populations are accurately known.

— The Red-billed Quelea, a tiny African waxbill finch (family Estrildidae), may be the most abundant wild, native bird species in the world. It occurs in locust-like swarms in the dry savannas south of the Sahara, where it wreaks havoc on grain crops. Single invasions of a particular region have been estimated to be 100 million birds strong, and the total population has been estimated at about 10 billion.

— The European Starling and the House Sparrow have been nominated as the world's most abundant land birds by some authorities, but others point out that while their distribution is wide, their occurrence is relatively localized.

— Estimates of the world's total domestic chicken population range from around 19 billion to more than

50 billion as of 2020 as compared with the total current human population of 7.8 billion.
— The most abundant species of wild North American bird known was the (now extinct) Passenger Pigeon; about 3 billion are thought to have been alive at the time of Columbus.
— Based on its extremely broad distribution, local population densities, and vast wintering roosts containing millions of birds, the Red-winged Blackbird has been judged the most numerous of living native North American land birds.
— The Red-eyed Vireo may be the most abundant of eastern North American deciduous-woodland birds. Of course, any one of many land-bird species with high population density and broad distribution can enter this contest without fear of being too badly defeated.

Air Conditioning (Do birds sweat?)

Birds do not perspire by means of sweat glands in the skin as humans do, so they must avoid overheating by other means. To some degree their air-conditioning system is simply the reverse of their heating system. Instead of "fluffing" feathers to increase insulation, for example, they compress their plumage to retain as little body heat as possible. Or they can increase circulation to unfeathered parts to give off more heat rather than reduce circulation to retain it.

Many Temperate Zone species molt into a thinner plumage for the summer or achieve one gradually by normal feather loss and/or wear. Birds also adapt their behavior to the air temperature, often seeking rest and shade during the hottest part of the day.

Though they cannot perspire, birds do vaporize water in the lungs and internal air sacs and release it by panting; this explains why you can often see many types of birds holding their bills open on hot days. Another cooling device is known in ornithological jargon as "gular flutter." The bare skin of the throat is quivered rapidly while the blood flow to the region is increased, thus letting off internal heat. This practice is especially prominent in birds with fleshy throat pouches, for example, cormorants, frigatebirds, pelicans, and boobies. A good place to witness either or both of these phenomena is at a seabird colony at midday when parent birds must sit on their nests and endure the full force of the sun while protecting their eggs or young from it.

By human standards, practicality has always characterized the species of American vultures far better than fastidiousness. Their method of cooling off is consistent with their reputation—they defecate on their feet. Species of storks—now known to be closely related to the New World vultures—also practice this untidy method of evaporative cooling.

Ali, Sàlim (1896–1987)

Creator of an environment for conservation in India, your work over fifty years in acquainting Indians with the natural riches of the subcontinent has been instrumental in the promotion of protection, the setting up of parks and reserves, and, indeed, the awakening of conscience in all circles from the government to the simplest village. Since the writing of your book, the Book of Indian Birds, your name has been the single one known throughout the length and breadth of your own country, Pakistan, and

Bangladesh as the father of conservation and the fount of knowledge on birds.

—FROM THE TRIBUTE TO SÀLIM ALI ON THE PRESENTATION BY THE WORLD WILDLIFE FUND OF THE SECOND J. PAUL GETTY WILDLIFE CONSERVATION PRIZE, FEBRUARY 19, 1976

The above commendation is a fair description of the impact that Sàlim Ali had on conservation, education, and advocacy on behalf of birds in India and beyond, but it fails to capture the richness of his ornithological journey over 80 years. What follows aims to fill in just a few of the details.

Sàlim Moizuddin Abdul Ali, "the Birdman of India," was born into a well-to-do Muslim family in Bombay, the youngest of nine children. His earliest interest in birds was as targets for "sport shooting" with his air gun, until one W. S. Millard, secretary of the Bombay Natural History Society, noticed that one of his victims was an unusual species, a Yellow-throated Sparrow. (The encounter is described in Ali's autobiography *The Fall of a Sparrow*.) This got the boy a tour of the Society's collection of mounted birds, and by age 10 he had started to keep a journal on his bird observations.

Ali's early academic career was "spotty," barely qualifying for Bombay University in 1917, then dropping out to look out for some family business interests in Burma (tungsten mines and timber), which allowed him to nurture his natural history interest (and shooting skills) in the nearby forests. Returning to school, he briefly studied business law but was persuaded by a perceptive professor to switch to zoology.

He married in 1918, and his wife, Tehmina, became a devoted companion in his future travels; they had no children, and he was devastated when she died in 1939 following a minor surgery. He never remarried.

Passed over for an ornithologist position at the Zoological Society of India in 1928, Ali decided to continue his studies in Germany at the Berlin Natural History Museum, where he worked for the eminent ornithologist Erwin Stresemann, who, in Ali's words, became his guru and provided entry into the world of international ornithology where he interacted with other prominent "birdmen" such as Ernst Mayr, S. Dillon Ripley, and Colonel Richard Meinertzhagen.

Ali was happiest when he was doing field studies and faunal surveys, which he pursued throughout India and had little taste for the systematics and taxonomy that obsessed many ornithologists at the time. In 1956, he wrote to Ripley, "My head reels at all these nomenclatural metaphysics! I feel strongly like retiring from ornithology if this is the stuff, and spending the rest of my days in the peace of the wilderness with birds, and away from the dust and frenzy of taxonomical warfare." He later said that he was dedicated to studying "the living bird in its natural environment."

For his part, Mayr complained to Ripley that Ali didn't collect enough birds to make the series of specimens necessary for taxonomic analysis. Contrariwise, Meinertzhagen, who was accompanied by Ali on an expedition to Afghanistan in 1937, complained that "he is quite useless at anything but collecting," and added this interesting "cultural" comment which reflects as much on Meinertzhagen as on Ali: "Sàlim is the personification of the educated Indian. . . . He is excellent at his own theoretical subjects, but has no practical ability. . . . His views are astounding. He is prepared to turn the British out of India tomorrow and govern the country himself."

Ali eventually found his way both as a scientific and popular writer about birds and as an effective and tireless advocate for conservation. He coauthored the authoritative 10-volume *Birds of India and Pakistan* with Dillon Ripley as well as the immensely popular *Book of Indian Birds* and many other publications. He was instrumental in the establishment of the internationally significant wetland reserve in Rajasthan alternately known as the Bharatpur Bird Sanctuary or Keoladeo National Park and in preventing the destruction of what is now Silent Valley National Park in the Nilgiri Hills, Kerala. Having maintained his relationship since childhood with the Bombay Natural History Society, he was able to play a key role in supporting it when it fell on hard times in the 1940s by appealing to Prime Ministers Jawaharlal Nehru and Indira Gandhi (who was an avid birdwatcher!).

Space does not permit noting more than a sampling of all the honors that Sàlim Ali accrued by the end of his long career. He was the first non-British citizen to receive the Gold Medal of the British Ornithologists' Union; and was also awarded the John C. Phillips medal from the International Union for Conservation of Nature and Natural Resources; and was made Commander of the Netherlands Order of the Golden Ark by Prince Bernhard, as well as receiving two of India's highest civilian honors and the Getty prize referenced above. In addition, his name graces Sàlim Ali's Swift (*Apus salimalii*) and several bird subspecies—as well as one of the world's rarest bats (*Latidens salimalii*)!

Altitude

We are impressed (as we should be) when we watch an eagle disappear from view thousands of feet above us and when we hear of tiny songbird migrants routinely traveling in the night sky at 5,000 to 10,000 feet (see MIGRATION). Yet many small birds that live above the tree line on the world's highest mountains double or triple these altitudes as they flit from one lichen-covered rock to another. More impressive still are the record holders listed below.

— 37,100 feet (11,300 meters), the highest documented bird flight: A Rueppell's Griffon (a species of large vulture) that collided with an airplane over the Ivory Coast (West Africa) on November 29, 1973

— 33,000 feet (10,000 meters): Common Cranes migrating above the Himalayas

— Approximately 29,000 feet (8,800 meters): A flock of migrating Bar-headed Geese heard calling above the Himalayan peak Makalu, elevation 27,766 feet

— 27,000+ feet (8,200 meters): A flock of Whooper Swans reported by a commercial airline pilot over the Outer Hebrides, Scotland, December 9, 1967

— 26,000 feet: Alpine Choughs; highest known altitude for a songbird recorded by the British expedition to Mount Everest in 1924. These members of the crow family followed the climbers' camp as it ascended, probably scavenging leftovers, as is typical of this species, which habitually forages above the tree line.

— 21,000 feet: Highest recorded altitude for small land bird migrants, detected by radar over Norfolk, England.

No human being can engage in strenuous exercise above 20,000 feet without supplementary oxygen. But these avian high fliers have the ability to hyperventilate when circumstances demand it, increasing oxygenation of the blood and the availability of oxygen to the lungs. They apparently also have an altered structure in the protein hemoglobin that increases its already considerable affinity for oxygen.

Anting

The application by birds of the body fluids of ants (or other substances) to their plumage. In "passive anting" a bird squats on an ant nest, and allows the insects to crawl over its plumage, sometimes while anting "actively" or at least going through the motions. In "active anting," a bird crushes an ant in its bill and rubs it vigorously on parts of its plumage, particularly the underside of the wing tips and tails coverts (feathers that lie at the base of the tail, both above and below). The ant fluid may be spread on the head and other parts from the wing tips, but certain parts of the body are treated only infrequently. More than 200 species of birds have been observed anting worldwide. It is practiced chiefly by songbirds.

The purpose of anting is not fully understood, but some credible hypotheses have been offered based on critical observations. Ants that emit chemicals (especially formic acid) as a defense technique are favored over ants whose main defense is stinging. It has also been demonstrated that formic acid kills feather mites (see ECTOPARASITE). The consensus theory is that anting is an avian form of delousing. The activity is often followed by PREENING and bathing).

In the absence of ants, birds will use mothballs, citrus fruits, vinegar, still-glowing embers, and other substances, most of which, like formic acid, produce a burning sensation. This kind of stimulation apparently "releases" the anting impulse, and since individual birds must discover the effect for themselves, some members of a species ant regularly, while others fail to learn the behavior.

An alarming side effect of certain birds' fondness for a burning sensation is a propensity for picking up flaming tinder and transporting elsewhere—in some cases to nests constructed in buildings, sometimes with disastrous consequences.

Apocalypse

We are presently witnessing a global extinction of species on an unprecedented scale and at an increasing rate. The causes, now abundantly documented, include: the rapid and widespread destruction of the planet's tropical rainforests (which contain more than 50% of the world's living organisms despite covering only 6% of the Earth's surface) and other habitats, mainly due to agricultural conversion; pervasive contamination of our air and water with toxic chemicals; and the destructive effects of CLIMATE CHANGE—to name just three forms of environmental degradation caused directly by our species.

Birds continue to play a heroic role as "canaries in the (global) coal mine," demonstrating by their conspicuous decline the extent of our peril. For example, according to BirdLife International, 279 bird species and subspecies have gone extinct globally in the last 500 years, and as of 2018, 40% of the world's 11,000 bird species are in decline and 1 in 8 species is threatened with extinction. Most alarming are recent comprehensive studies conclusively documenting the recent and continuing decimation of the majority of bird *populations*. According to research published in the journal *Science* in 2019, numbers of individual birds in Canada and the United States have fallen by 29% since 1970—that is, there are now nearly 3 billion fewer birds inhabiting North America than there were just 50 years ago. The decline affects many species such as robins, jays, and blackbirds, normally considered abundant. And iconic species such as Snowy Owl and Atlantic Puffin are officially deemed "Vulnerable" and may be on a path to extinction.

In retrospect, it hardly seems surprising that we are now in the throes of a "bird apocalypse," considering the myriad means of extermination that we have practiced, as briefly summarized below.

OVERHARVEST. Unregulated killing for commercial markets as well as for sport, which went unchecked until the early twentieth century, can be credited with the extinction of once-abundant species such as the Passenger Pigeon and the Great Auk. It also drastically reduced the numbers of many shorebirds/waders and upland game birds before effective conservation laws were passed. Egrets were decimated by plume hunters between about 1875 and 1900.

HABITAT DESTRUCTION. This has long been considered the single greatest threat to birdlife on the planet, though chemical pollution is presently giving it a run for its money (see below). It occurs in direct proportion to population growth and development, though it takes many forms: draining and filling of wetlands for highway and mall construction; destroying natural grasslands for agriculture; clearing of forests for lumber and house lots.

Habitat destruction continues to outstrip conservation. The clearing of tropical rainforest was estimated in 1989 to be continuing at the rate of about a football field per second. Since then there have been fleeting hopes of increased protection, but under present (2020) anti-environmental/pro-business and autocratic political leadership, the scope and rate of the forest destruction is greater than ever. Furthermore, we now know that the survival of tropical forests is crucial to such fundamental matters as the planet's oxygen supply, its climate, and its genetic diversity.

PESTICIDES AND OTHER CHEMICAL POISONS. The indiscriminate use of highly toxic chemicals to control insect pests stands as one of the chief follies of the modern era. Those who continue to urge heavy reliance on nonspecific, highly toxic, persistent pesticides argue that they are the most effective and efficient means of controlling pest insects, such as mosquitoes, caterpillars, and leaf beetles. Opponents reply that such pesticides:

— Are insufficiently specific, killing "good" insects as well as "bad" along with other insect-controlling organisms such as birds and amphibians, and potentially destabilizing entire ecosystems.

— Are highly persistent and mobile both in soil, whence they are washed into bodies of water, and in animal tissue, where they accumulate and increase in concentration as they are passed up a food chain. Pesticide residues have been detected in the tissue of humans living in the remotest reaches of the Earth.

— Exacerbate the problem since their effect is short term; insect populations readily produce resistant generations, which in turn require the use of even deadlier pesticides or heavier applications of the original one.

— Continue to be developed and promoted in ever more destructive forms such as neonicotinoids, a family of pesticides that has proven to be highly destructive of the insect populations on which many birds depend. The recent documented decline of aerial and other insect populations has been described as an "insect apocalypse" that is implicated directly in the bird apocalypse.

LEAD POISONING is one of the oldest and most pervasive man-made threats to gallinaceous birds and waterfowl, which ingest, along with other grit, pellets from shotgun cartridges—the inevitable remains of the thousands of rounds of ammunition spent during hunting seasons. In addition, loons have been found to suffer high mortalities caused by the ingestion of lead fishing sinkers, and raptors are often killed by shot ingested when feeding on unretrieved game. The use of lead shot for hunting ducks, geese, and swans is now banned by federal law in the U.S., though of course, tons of lead shot remain in wetland sediments.

OIL SPILLS. It does not require much imagination to understand that petroleum, seawater, and swimming birds do not mix well. Loons, grebes, tubenoses, sea ducks, alcids, and other oceangoing species that are caught in the thick phlegm of an oil spill quickly become hopelessly mired, totally unable to function, and are thus doomed to a slow death by starvation, exposure, and poison. Even birds with apparently negligible oil stains may be dangerously liable to total loss of body heat, because the oil creates a gap in the insulation system that maintains a bird's body temperature on the cold ocean.

COMPANION ANIMALS. The dogs, cats, rats, pigs, goats, and other animals that have followed the spread of Western civilization around the world pose serious threats to birds and other wildlife. Their arrival has been particularly devastating among endemic island species such as the (now extinct) Laysan Rail, which, due to the absence of predators on its native island, had lost the ability to fly. However, any burrow-nesting or ground-nesting bird is potentially threatened by these animals. Cats are especially devastating to bird populations. It is estimated that there are presently about 60 million feral cats at large in the U.S. together with 60–88 million owned cats, a large percentage of which are allowed to run free. Cats have been responsible for the global extinction of 33 island-inhabiting species worldwide, and current science has confirmed that cats kill at least 1.3–4.0 *billion* birds a year in the lower 48 states.

MAN-MADE STRUCTURES. Skyscrapers, power-generating windmills, and television towers represent hazards to low-flying migrants and other birds. Especially on foggy nights, the bright lights of city buildings may attract thousands of birds, which become disoriented, unable to find their way out of the maze of reflecting glass, and often die by flying into windows in their panic. Johnston and Haines (1957) record two par-

Bird Enemy #1

ticularly disastrous nights in early October 1954 during which more than 100,000 migrant birds were killed in only 25 localities sampled from New York to Georgia. A reasonable estimate of birds killed annually in North America by encounters with buildings, picture windows, and TV towers begins in the millions.

AUTOMOBILES. In the 1960s it was estimated that about 2.5 million birds were killed annually in collisions with automobiles in Britain. The number of birds killed on North America's much more extensive network of highways must be staggering.

INVASIVE EXOTICS. While introduced bird species such as European Starlings and House Sparrows have caused population declines in native cavity-nesting birds through competition for nest sites, a more insidious threat is that from invasive plant species. A striking instance is that of native marsh bird communities, including species such as grebes, bitterns, and rails, many of which are becoming scarce due to wetland degradation caused in significant measure by invasive exotic plants such as Common Reed and Purple Loosestrife. Eradication of these plants once they are established has proven to be labor intensive, expensive, and sometimes impossible.

THREATENING THE WHOLE. One reason so many people have recently become interested in the welfare of the world's wildlife is that our own fate is inextricably involved with that of our fellow organisms.

Listed above are a few of the many threats to the environment that we have posed quite recently. They are ones that relate particularly to birdlife and in ways that have been clearly documented. Many others, such as air pollution, acid precipitation, disposal of chemical and nuclear wastes, and the release of greenhouse gases leading to global CLIMATE CHANGE—both alone and in combination—gnaw inexorably away at the abundance and diversity of earthly life. This self-inflicted corrosion afflicts not only our physical well-being but also what we alone have the ability to appreciate—the overall quality of life.

It was once traditional among coal miners to take a caged canary down into the shaft with them. When the canary fell off its perch—killed by lethal fumes—it was time to get out. If we wake up one morning to find birds falling off their perches, it will be too late to correct earlier miscalculations. And, of course, there is nowhere else to go.

See also MORTALITY.

Audubon, John James (ca. 1785–1851)

The name the world associates inextricably with birds belonged to an artistic genius who, like many of his kind, was as eccentric as he was brilliant. He was born at Les Cayes, Sainte Dominique (now Haiti), the acknowledged bastard of a well-to-do French merchant-slaver, Jean Audubon. The artist once described the mother he never knew as being "a lady of Spanish extraction . . . as beautiful as she was wealthy," but this appears to be an example of the autobiographical imagination that encouraged him to let it be rumored that he might be the lost Dauphin of France. His mother seems to have been Jeanne Rabin, a French chambermaid on his father's Caribbean estate, since Audubon was originally called Jean Rabin. But his father soon adopted him (along with a daughter also born out of wedlock), removed him to his French estate and rechristened him Jean Jacques Audubon.

Having escaped from slave riots and revolution in his birthplace at age 3, Audubon arrived in France in the midst of the Reign of Terror, and when he was 17, his father sent him to his farm outside of Philadelphia to avoid Napoleon's draft. By this time Audubon had acquired the education and graces of a well-to-do French citizen, including competence in riding, fencing, playing the violin—and drawing. His father also encouraged his interest in natural history, especially birds.

Like many another genius, Audubon was a failure at workaday occupations such as family and business; even his early labors as an itinerant portraitist amounted to little except, probably, to hone his drafting skills. He lost his last business as a result of a banking crisis and

After Audubon's self-portrait
(detail in Golden Eagle painting for *Birds of America*)

a trade embargo imposed during the War of 1812. His response to bankruptcy was to leave his (very capable—and long suffering!) wife Lucy to raise their two sons, while he joined the crew of a river flatboat and headed into the wilderness to fulfill his dream to paint every bird species in America. He had never lost his interest in the natural world, and he wrote in his journal at this time: "*I never for a day gave up listening to the songs of our birds, or watching their peculiar habits, or delineating them in the best way that I could.*"

To the world's immeasurable benefit, his *raison d'etre*, the double elephant folio edition of *The Birds of America*, consumed Audubon body and soul until its completion in 1838. These plates, engraved largely by Robert Havell, Jr., of London from Audubon's original watercolors and hand-colored by artisans under the artist's direction, comprise 497 bird forms (many of Audubon's "species" have been "lumped"; see LUMPING) in most cases painted life-size. They also contain a wealth of carefully rendered plants (especially those by the teen-aged Joseph Mason), as well as insects and other life-forms by Maria Bachman, and a fine series of period landscapes and townscapes.

As was common practice at the time, *The Birds of America* was sold by subscription and published serially. Thus, while between 175 and 200 complete, bound sets of the 27 × 40-inch double-elephant folio edition are thought to have once existed, an unknown number of incomplete sets were also published. Many of the complete sets have since been broken up and the individual leaves sold for increasingly impressive sums. The original subscribers paid $1,000 for the four huge volumes containing 435 plates—a grand enough sum in the 1830s. However, today a single plate can command as much as $200,000, and a complete elephant folio sold at auction for US$10.27 million in 2010 and another for US$9.65 million in 2018 (up from $440,000 in the early 1980s).

After completing *The Birds of America*, which brought its creator fame and a measure of financial security, Audubon embarked in 1859 on a smaller, octavo edition in collaboration with the Philadelphia lithographer

J. T. Bowen and a full-sized chromolithograph edition by Julius Bien, which due to the advent of the Civil War and unscrupulous business practices was never completed.

Also begun right after the publication of the elephant folio was Audubon's collaboration with his friend the Reverend John Bachman on *The Viviparous Quadrupeds of North America.* But the master's eyesight and energy were beginning to fail, and many of the paintings for this work were done by his son, John Woodhouse Audubon. He grew senile in the last two years of his life and died at Minniesland, his home overlooking the Hudson in upper Manhattan.

Though Audubon represented himself abroad as a colorful frontier character, "an American woodsman," acting and dressing the part with long flowing tresses and rustic, fur-trimmed attire, he was not the naive, rude "natural genius" he sometimes pretended to be. He had studied in Paris before coming to his father's farm in Pennsylvania at the age of 17, and his shrewdness in promoting himself and his paintings in the highest circles of European society is verified by the publicity he quickly generated and his success at peddling such an expensive work. His wit and sophisticated knowledge of world art traditions, as well as of bird behavior, came through in many of his dramatic compositions and in his writings. Both his Ornithological Biographies, written (with William MacGillivray) to accompany *The Birds of America*, and his journals, edited by his granddaughter, Maria A. Audubon, contain many passages that are movingly evocative of the early nineteenth-century American wilderness and its inhabitants. Audubon was among the first to lament their sad destruction.

Bailey, Florence Merriam (1863–1948)

Bailey, Florence Merriam (1863–1948)
According to her friends and admirers, the ornithological career of Florence Merriam Bailey owes its great distinction to a combination of personal traits, including a strong affection for the natural world, keen powers of observation, respect for science, and a lucid yet charming writing style.

Florence Augusta Merriam was born and spent her childhood at her family's estate on a wooded hilltop at the edge of the Adirondacks. She and her brother, C. Hart Merriam, who also became a distinguished naturalist, were encouraged by both parents to interest themselves in the natural world from an early age. And she benefited from her well-to-do family's "connections": Her father corresponded with John Muir after meeting him on a trip to Yosemite, and Ernest Thompson Seton was an early mentor and encouraged her interest in birds. After prep school she attended Smith College, and though she did not receive a degree, it was during this time that she became an activist in the bird conservation movement. Protests against decorating ladies' hats with wild bird plumes were gathering steam, and Merriam founded chapters of the Audubon Society at Smith and later in Washington, D.C., where she worked to pass legislation to end the plume trade. (For details of this movement, led almost exclusively by women, see HEMENWAY.)

Bailey published a number of important books on birds for different audiences. Her *Birds Through an Opera-Glass* (1890), a beginner's manual covering 70 species, has been called the first modern field guide; she followed up with *Birds of Village and Field* (1898), covering 150 species. In between she published *A-Birding on a Bronco* (1896), an account of her time spent visiting her uncle in Southern California—partly to treat her likely diagnosis of tuberculosis—and exploring the local birdlife; it was the first book to be illustrated by the soon-to-be eminent Louis Agassiz Fuertes.

In 1899, she married Vernon Bailey, a naturalist and mammalogist who became chief field naturalist for the U.S. Bureau of Biological Survey. Though they made their permanent home in Washington, D.C., the couple spent the next three decades doing fieldwork throughout the American West, but especially in what was then the Territory of New Mexico and in Arizona. From this experience emerged her most substantive ornithological works, including the *Handbook of the Birds of the Western U.S.* (1902), complementing Frank Chapman's previously published eastern handbook. From youth Bailey had been interested in all aspects of birds' behavior, not only their identification, distribution, and taxonomy, which occupied most professional ornithologists of the time. These interests were amply represented in the *Handbook*, which, however also demonstrated Bailey's deep immersion in the relevant literature and plumage details from poring over thousands of specimens in the collections of the Smithsonian. With more than 600 illustrations, this was no beginner's book but one of the most authoritative ornithological manuals to be published to that time.

In 1916, following the untimely death of Welles Cooke, who had begun a study of the birds of New Mexico, Bailey was asked to complete his project. Instead she turned it into her own magnum opus based on her extensive fieldwork as well as on Cooke's notes. The result of a 12-year effort was *Birds of New Mexico* (1928), a comprehensive work comprising more than 700 pages of text, 22 full-page color plates by Allan Brooks, 34 black-and-white illustrations by Louis Agassiz Fuertes, 60 range maps, and 29 pages of literature citations.

Bailey's life's work was honored by her election as the first associate member of the American Ornithologists' Union, and *Birds of New Mexico* made her the first female recipient of the Brewster Medal, awarded for "an exceptional body of work on Western Hemisphere birds."

Baird, Spencer Fullerton (1823–1887)

A central figure in American ornithology in the latter half of the nineteenth century and, perhaps more than any other single person, responsible for bringing knowledge of the North American avifauna from the pioneering stages of WILSON and AUDUBON to the comprehensiveness and consistent classification of the modern era. In the Historical Preface to his *Key to North American Birds* (1890), Elliott COUES outlines the "Bairdian Epoch" in American ornithology, and at Baird's death, the ornithologist J. A. Allen called him the "Nestor of American Ornithology."

Baird started out as a boy naturalist in Pennsylvania, took his first degree in natural history at age 17 and a Masters shortly after from Dickenson College, which then gave him a professorship. He made friends before he was 20 with the aging Audubon, and he rapidly developed talents as a linguist, administrator, politician, and writer. He published extensively not only on birds but also on mammals, reptiles, and fishes. He became the secretary of the Smithsonian Institution in Washington, D.C., and persuaded Congress to build the National Museum of Natural History to house the collections he accumulated so assiduously. He created and headed the U.S. Commission of Fish and Fisheries and founded the marine science laboratory at Woods Hole, Massachusetts. Through his connections in the government and his relationship by marriage (son-in-law) to the Inspector General of the Army, he was responsible for the remarkable involvement of government surgeons in the collecting of zoological specimens in western North America. These collections along with his own fieldwork provided the basis for a government report (part of the survey for the railroad route to the Pacific) on birds of the United States. Later reissued as *Birds of North America* (1859), this work included 738 forms, constituting the first truly comprehensive, scientifically organized checklist of the avifauna. In addition to his achievements, he was remembered as a great friend and supporter of other naturalists. His achievements are commemorated in the scientific and standard English names of Baird's Sandpiper (*Calidris bairdii*) and Baird's Sparrow (*Centronyx bairdii*).

Bergmann's Rule

The observation (formulated by a nineteenth-century German zoologist) that overall body size tends to be greater in representatives of bird and mammal species living permanently in cooler climates than in those living in warmer climates. Put simply, American Kestrels in Maine (*Falco sparverius sparverius*) are larger on the average than those in Florida (*F. s. paulus*). The ecological principle followed is that large bodies retain heat more efficiently than small ones, and that there is therefore an adaptive size increase in populations constantly subjected to lower temperatures. In 1983, ornithologist Frances James switched eggs of Florida Red-winged Blackbirds into Minnesota nests and vice-versa. Despite their origins, the changelings ended up larger in the North and smaller in the South, suggesting an environmental, rather than a genetic basis for Bergmann's rule.

Bill

A bird's bill (less formally "beak") is analogous to our jaws, though it is used differently in many respects and bears little obvious resemblance in form. It consists of two bony frontal extensions of the skull—one above and one below the mouth—covered with a horny or leathery sheath made of keratin, which, like our fingernails, is a product of the upper layer of the skin (epidermis). In most vertebrates the upper jaw is called the maxilla and the lower jaw the mandible, but in birds the two parts of the bill are often known as the upper and lower mandibles. Both mandibles have two sharp edges, sometimes equipped with some form of serration for holding food securely. In most birds the upper mandible is slightly longer, deeper, and wider than the lower mandible, so that there is a slight overlap from above. Nostril holes are evident in all but a few species and typically occur nearer the base of the bill than the tip. Bill wear is compensated for by formation of new keratin much in the way our outer skin layer flakes away and is imperceptibly replaced, and broken bills can also "grow out" in some cases.

LENGTH. Mandibles, like fingernails, grow continuously and are worn down to their characteristic size and shape through constant use. Bills vary greatly in length from a few millimeters in swifts to a foot and a half in the largest male pelicans. The longest bill in the world is that of the Great White Pelican (*Pelecanus onocrotalus*) at 18.5 inches (47.1 cm). The largest Old World storks are close runners-up, several species bearing beaks in the 13.5-inch (35 cm) range. The bill of the Long-billed Curlew, which sometimes appears almost grotesque in proportion to its owner's body, actually measures "only" 8.6 inches at its longest (maximum females), the difference of course is a matter of proportion and perception.

SHAPE. Bills also come in a wide range of shapes adapted to various feeding habits. In addition to obvious adjectives such as "short," "long," "straight," "curved," "broad," "narrow," and "pointed," a number of specialized terms are used to describe particular bill types, for example, *recurved* (i.e., curved upward as in the avocets); *decurved* (i.e., curved downward as in the ibises and curlews); *serrate* (i.e., with toothlike modifications along the edges of the bill); and many others. See below for examples of highly specialized bills.

USE. Because a bird's bill is inevitably associated with its mouth, and because birds spend a great deal of time eating, it is natural to associate bill functions with food. Perhaps, therefore, it should be emphasized that many of the things birds do with their bills—including catching and preparing food, but also carrying, digging, nest building, and defense—are things we would do with our fingers and hands. The most generalized bill function is a pliers-like grasping, performed to some extent by all birds. However, the diversity of uses to which birds put this essential appendage is limited only by the imagination. A partial list might include tearing (of food items), probing, stabbing, hewing (of nest cavities), husking (of nuts and seeds), straining, and display.

A few highly specialized bills deserve additional description:

KIWIS. All five species of these remarkable birds—endemic to the islands of New Zealand—have very long bills, which are unique in having the nostrils located at the tip of the bill. They also have an exceptionally well-developed sense of smell. These two traits allow kiwis, feeding at night, to probe the earth and smell and capture worms and other prey without having to see or feel them.

SPOONBILLS have long bills that are strongly flattened vertically and widened into a spatulate shape at the tip. The interior of the "spoon" is lined with sensitized tissue that feels small food items as the birds swing their partially opened bills through shallow, murky water. When small prey such as insect larvae or shrimp are encountered, the "spoon" snaps shut on them.

OYSTERCATCHERS' bills are laterally flattened toward the tip. This thin, sharp tool is inserted chisel-like between the shells of live bivalves (not just oysters) and then shears the powerful adductor muscle(s), which otherwise hold the shells closed "as tight as a clam." Anyone who has tried to open an oyster or hard-shelled clam without proper tools will appreciate the finesse in this evolutionary achievement.

Black Skimmer

SKIMMERS are unique in having a lower mandible that is significantly longer than the upper—an adaptation to an equally unique method of feeding. The birds fly low over calm waters (often at dawn and dusk and in flocks) and "shear" the surface with the tip of the lower mandible, which is laterally flattened to nearly razor sharpness. When a fish or other edible object is encountered by the blade, the upper mandible closes down on it instantaneously.

GROSBEAKS and certain other finches have massive bills especially adapted to cracking very hard seeds. The size of the bill, the presence of specialized husking mechanisms on the inside of the bill, and the strength of the jaw muscles determine how much force can be brought to bear upon a seed. The champion in this category appears to be the Eurasian Hawfinch, which can exert a pressure of up to 100 pounds per square inch, though the nutcracking ability of other species such as the North American Evening Grosbeak may be comparable.

CROSSBILLS. The seemingly deformed bills of these finches are in fact sophisticated tools for extracting seeds from tightly closed conifer cones. The tip of the longer upper mandible (which curves downward) is wedged between two of the cone scales so that the curve of the lower mandible rests on the surface of the cone and the tip of the upper mandible is fixed on the inside surface of a scale. The bird then twists its head in such a way that the force exerted against the stable lower mandible forces the scale apart on the tip of the upper mandible. Simultaneously, the two mandibles are forced apart by special jaw muscles—not in the usual vertical plane of birds opening their bills, but in a horizontal (lateral) motion that contributes essential force to the separating process. Once the scales have been forced apart, the unusually large, protrusible tongue reaches into the cavity and detaches the firmly anchored, unripe seed with the aid of a special cartilaginous cutting tool that forms the tongue tip.

Birding while Black

In the United States, Black birdwatchers have historically been *rarae aves*. In 1995, American birders accounted for about 24% of the white population and 86%–88% of all birders, then as now, self-identified as white. Since 1995 the percentage of Hispanic birders has increased from 1.9% of total Hispanics in 1995 to 10.8% by 2001. While this is good news for multicultural improvements in what has sometimes been characterized as an elitist pastime, it does not much improve the demographics for Black bird lovers. Between 1995 and 2006, the percentage of birders among African Americans hovered between 6% and 8.2% in the population as a whole. While there has been some improvement in these numbers since, the upward trend has not been dramatic. As author, poet, and PhD wildlife biologist J. Drew Lanham has pointed out in his 2016 depiction in *The Homeplace: Memoirs of a Colored Man's Love Affair with Nature,* "The chances of seeing someone who looks like me on the trail are only slightly greater than that of sighting an Ivory-billed Woodpecker."

That people have recently experienced a "Who knew? moment" about Black birdwatchers is strongly attributable to the Memorial Day (2020) contretemps in New York City's Central Park between Harvard-educated senior biomedical editor (and former Marvel comics writer) Christian Cooper (a well-known Black birdwatcher) and a white woman, Amy Cooper (no relation) who had illegally let her dog off leash. When Mr. Cooper politely requested that she obey park rules and leash her pup, she became belligerent and then called the police saying that she was being threatened by an African American man. A video of the encounter went viral. The good news resulting from this incident is (1) that the woman was broadly condemned for her behavior; and (2) that the Black birding community has gained increased visibility—possibly leading to an increase in participation. (?)

An immediate response was the declaration of the first Black Birders Week, May 31–June 5, 2020, a series of virtual events run on Twitter and Instagram with the following goals:

1. To increase visibility and representation of Black birders, naturalists, and explorers;
2. To spark a necessary dialogue about the very real threats and racism experienced by Black birders; and
3. To push institutions to go beyond diversity programs and implement inclusion efforts that give Black birders a space to be seen and heard.

The event was immediately embraced by the National Audubon Society, the American Birding Association, the National Wildlife Federation, and many other organizations and government agencies.

To view personal accounts of the joys and hazards of birding while Black, tune into activist-birder Jason Ward's YouTube series *Birds of North America*.

Birdwatching

Usually refers to the regular, somewhat methodical seeking out and observation of birds, whether for pure aesthetic pleasure or for recreation or out of a more serious, scientific motive. That is, the term usually does not apply either to people who may feed birds but are not particularly interested in which species they attract, or to professional ornithologists, most of whom spend at least as much (usually much more) time in the laboratory and library as they do observing birds in the field. The relatively new but rapidly evolving behavioral branch of ornithology intrinsically involves long hours of watching birds, yet the practitioners of this discipline are known as bird behaviorists, or behavioral ecologists, implying a specialized education that birdwatchers usually lack.

The semantics, history, and social implications of "birdwatcher"/"birder"/"ornithologist" are rather complex and sometimes amusing. It should be noted first that many serious birdwatchers of today could easily instruct an ornithologist of 100 years ago—due, of course, to the accumulation of knowledge that the early ornithologists helped to discover. Nevertheless, modern "birders" should neither feel nor be disparaged by modern ornithologists; their contributions have been significant and are increasing with the rise of Citizen Science.

In North America an interest in birds has traditionally been associated with somewhat different values than it has in Europe and especially Britain, where birds and man meet on a more nearly equal footing than in any other country. Until after World War II, birdwatching in America was regarded as a pursuit of (1) the rich (perhaps because many early conservationists, e.g., Teddy Roosevelt, were American aristocrats); (2) the eccentric (because of field clothes? because only a crazy would spend time peering up into trees at tiny, flitting forms?); and in males (3) the somewhat effeminate (not enough physical contact for a proper sport!). Of course these were and are largely stereotypes pervasive in popular culture, but hardly representative of today's birders, who are just as "normal" (and eccentric) as the population at large.

Yet despite the dramatic increase in environmental education and the popularity of environmental causes in the U.S. since the 1970s, there is still, in some quarters, some stigma felt by any boy older than 10 who would just as soon watch birds as play sports. In much of Europe, such negative values attached to an interest in natural history are not so apparent—the idea of a father and son going birdwatching in the morning and to a soccer match in the afternoon surprises no one.

Another interesting difference between American and European birdwatchers involves how each perceives his/her avocation. In a book aimed at British birdwatchers, Alan Richards observed, "In the early 1950s ornithologists tended to believe that all birdwatching should yield some scientific return; in more recent years this attitude has disappeared and birdwatching is usually done for sheer pleasure and interest." With some notable exceptions the situation in America has been just the reverse. In the 1950s (and later) most ornithologists here expected nothing whatever of birdwatchers, and most birdwatchers expected nothing of scientific value of themselves—only fun. Only recently have young American birdwatchers in any numbers begun "watching" (rather than simply identifying and listing) birds and carrying notebooks into the field to record their observations. This trend has accelerated over the last 30 years, with the popularity of Breeding Bird Atlas and eBird projects, and academic ornithology now actively promotes contributions by birders. Furthermore, as has long been realized in Europe, effective bird conservation must involve a constituency of committed people if it is to have any political leverage, and engaging troops of volunteers in bird censuses and other useful efforts has proven to be an excellent means of promoting conservation action.

Then there is the matter of what one calls oneself and one's interest. Twenty years ago, British field ornithologists called themselves "birdwatchers," a no-nonsense, reasonably descriptive term, and disdained the Americanism "birder" as a bit frivolous or trendy. For their part, Americans have long thought "birdwatcher" rather stodgy and smacking of some of the unwholesome traits noted above. "Birder" by contrast seemed to imply a more serious, aggressive approach to the activity.

Someone wishing to emphasize transatlantic differences in birding styles might say that American birdwatching (*birding*) has typically been a form of competitive sport involving how many species of birds one can see in a day, year, life, state, or the world; a great deal of driving and talking are usually involved in a day's birding, and relatively little attention is paid to numbers, behavior, and the like. European birders (*birdwatchers*) have tended to concentrate more on honing field identification skills to the level of microscopy and to think of their activities as worthwhile observations; they tend to walk (and slog) more, write and sketch a great deal in notebooks (field marks, behavior, phenological data); are not very tolerant of idle conversation while "watching," and are usually suspicious of "twitchers" (i.e., listers). They consider good notes on a particular species better evidence of a good day in the field than a long species list. Obviously, there are distinct advantages to both these approaches to birdwatching and, as was probably inevitable, they seem to be merging into an activity of global popularity that is both fun and of conservation value. By the year 2001, "birder/birding" had become the accepted terms of art in the United Kingdom, while "birdwatcher/birdwatching" is uttered more frequently and without seeming embarrassment in the U.S. There are many more twitchers in Britain now and apparently fewer (or at least less mindless?) listers in the United States than previously. And keeping records and supporting bird conservation is, wholesomely, far more prevalent among American birders than was once the case.

For a particularly acerbic British expatriate's account of North American birders, see Jack Connor's *Season at the Point* (1991). Bill Oddie (1980) also gets off some good lines on the subject in a chapter of his *Little Black Bird Book* titled, "What am I? What are you?"

See also LISTING; TWITCHING.

SOME DEMOGRAPHICS (based largely on U.S. Fish and Wildlife Service surveys 2006–2017).

POPULARITY. Birding is near the top of the list of most popular outdoor recreational activities in North America and its popularity continues to grow. As of 2006, 48 million Americans self-identified as "birders"; in these surveys, a birder is someone who has taken at least one trip a mile or more away from home to observe birds or has closely observed and tried to identify birds around the home; those who just noticed birds or looked at birds in a zoo were not counted. Total U.S. birders thus tops household food gardeners (43 million) and golfers (24 million). Only recreational fishing has topped birding at 49 million in 2017. Many more Americans (70.4 million) identified themselves as "having an interest in birds."

GENDER. Birdwatching was once a strongly male-dominated activity, the most adept and ardent birders often beginning as obsessive preadolescent boys. In 1975, 78% of American Birding Association members were male; by 1994 the ratio was down to 65%. But today, in contrast to male-dominated recreational fishing and hunting, women now make up 54% of all U.S. birdwatchers.

AGE. In the U.S., older birders dominate, with the largest age segment in the 45–55+ range (52%). The segment with the lowest participation is the 16–24 range at 8%.

WEALTH. The old stereotype of the well-to-do birdwatcher is still valid to some extent. To successfully pursue a respectable life list, a certain minimum of leisure and mobility are necessary, and as of the period of these surveys 29% of birders earned $75,000 or more a year and 56% made $50,000 or more. It is interesting that this number is slightly down from 1994–1995, possibly indicating that the activity is becoming more appealing to the less affluent. Birders have become known to chambers of commerce as a desirable demographic that often patronizes high-end stores or services. The rise of birding festivals and trade magazines pitched at "bird merchants" is a response to the proliferation of birder-consumers. The professions are well represented among birders, particularly doctors, who have always figured prominently in American ornithology.

EDUCATION. As of 2006, more than 28% of birders held bachelor's degrees (against 21.5% in the general population), and 68% had a high school diploma or more. In 1994, the educational level of members of the American Birding Association included 98% with high school diplomas, 80% with bachelor's degrees, and 43% with a master's degree or PhD. An interesting trend in this category is that the percentage of birders without high school diplomas increased between 1995 and 2006 from 7.6% to 12%, while the college-educated group dropped from 34.7% to 28%.

ETHNICITY. White birders (those of European and Middle Eastern descent, but not including Hispanics) make up 24% of the white population in the U.S., but 88% of American birders identify as white. The percentage of Hispanic birders rose from 1.9% of their total U.S. population in 1995 to 8% in 2006. By comparison the percentage of Black birders has remained low at 6% of all Black Americans. See BIRDING WHILE BLACK. Seven percent of Asian Americans are birders.

GLOBAL DISTRIBUTION. While birdwatching has been pursued historically mainly in "developed" countries of the Temperate Zones—especially North America, northern and western Europe, Japan, South Africa, Australia, and New Zealand, it is growing steadily in the developing world, especially as ecotourism and conservation programs are creating new economic opportunities there.

OTHER STATISTICS

— 24.7 million people take trips every year to watch birds.
— 85% of American Birding Association members travel outside their state to bird; 49% do so outside their country.
— 70 million Americans feed birds at home.
— In 1997 there were about 70 birding festivals in North America.
— "Avid" birders spent more than $200 million per year on birding related travel, optics, and other related retail items (not including bird feeding).

Based on the above figures, it seems fair to say that birding is not only growing dramatically as a popular pastime but also appealing to a more diverse audience.

It has also become a major market segment. These trends are all the more striking in view of the fact that participation in other popular outdoor activities involving wildlife, for example, hunting and fishing, is actually declining.

Bonaparte, Charles Lucien (1803–1857)

A nephew of Napoleon I, the young Bonaparte tackled the description of the North American avifauna in his *American Ornithology* during an eight-year stay in America. He was well educated in science and is generally acknowledged to have been one of the foremost ornithologists of his time. His American reputation is eclipsed somewhat due to his appearance on the scene between the two giants of early American ornithology, WILSON and AUDUBON. He continued to study and write about birds and other zoological subjects on returning to Europe, but much of his later life was consumed by politics, especially the independence of Italy. Bonaparte's Gull (*Larus philadelphia*) nests in the boreal forests of Canada and Alaska, where, uncharacteristically for a gull, pairs build their nest of sticks in coniferous trees.

Brood Parasitism

Describes the laying of one or more eggs by one bird in the nest of a "host" pair (often of a different species), which rears the chick or chicks, often at the expense of some or all of its own young.

INTRASPECIFIC BROOD PARASITISM. Laying eggs in the nests of neighboring members of the same species is now known to be a common practice among songbirds and certain other groups of birds, including grebes, waterfowl, gallinaceous birds, gulls, and pigeons. It tends to be especially prevalent when population density is high (e.g., in colonial species) and when good nest sites are scarce. In some cases, as many as a quarter (colonial swallows) to three-quarters or more (ducks) of nests have been found to contain the eggs of another pair. European Starling females actually prospect for host nests of their own species and remove an egg from the clutch before laying a replacement. Because such parasitic eggs are virtually identical to those of the host, they are usually accepted (unless the receiving female has not yet laid any of her own eggs), and the unrelated young are reared along with the host's own. This model has little or no detrimental effect on the host's reproductive success and gives the parasite's genes an additional survival option. It has been suggested that such parasitism within species may have been an early stage in the evolution of *obligate brood parasitism* described below.

FACULTATIVE (DISCRETIONARY) BROOD PARASITISM describes cases in which birds occasionally lay eggs in nests of different (usually closely related) species. Several duck species practice this casual form of parasitism, and the Redhead and Ruddy Duck are known to avail themselves of the alternative regularly. In the species studied, the parasite's eggs typically outnumbered the host's own in completed clutches, conferring a reproductive disadvantage on the foster parents. Some land birds also parasitize each other occasionally, especially in years when good food supplies permit high egg production.

OBLIGATE BROOD PARASITISM refers to species that never build a nest of their own or care for their own eggs or young. It is characteristic of only about 1% of all bird species in 6 families worldwide, including one duck species; the majority of the 78 species of the Old World cuckoo family; 3 species of New World cuckoos; probably all 17 species of honeyguides; 5 of 6 cowbird species; one species of African weaver; and the whydahs and indigobirds in the genus *Vidua*.

Common Cuckoo chick (*L*)
being fed by its host,
a European Reed Warbler

BEHAVIORAL VARIATIONS. Brood parasitism has evolved independently in the families noted above, and its practice varies significantly among them. The Black-headed Duck of South America will use virtually any nest it can find (including that of at least one raptor species), does not displace eggs or young from the host's nest, takes no food from the host parents, and becomes fully independent within two days of hatching; it has therefore been called "the most perfect of avian parasites" (Weller 1968). Even more "virtuous" perhaps is the Giant Cowbird, whose nestlings, according to Smith (1968), actually benefit their host oropendolas in some circumstances by removing botfly larvae from their step-siblings. All the other obligate brood parasites inflict some degree of harm on their hosts: (1) by removal of host eggs or young (by parent or nestling parasites); (2) by consuming food intended for a nest's rightful inhabitants; and/or (3) by taking up inordinate amounts of the parent birds' time and energy.

The response of the host also varies, from (1) complete obliviousness of the deception and sacrifice of its own young to the needs of the (often much larger) parasite; (2) rearing the parasite along with some of its own young; (3) evicting the parasitic egg; (4) covering over a parasitized nest with successive "floors"; or (5) abandonment of the parasitized nest.

IDENTITY CRISIS? How does a young cowbird or cuckoo know it is not a warbler and how to find an appropriate mate? In cuckoos and cowbirds, the answer lies at least partially in their genetically inherited ability to sing and to recognize and respond to the songs of their own species. The African indigobirds, by contrast, adopt much of their host's identity throughout their lives. In addition to mimicking the appearance and behaviors of the host nestlings—species of firefinches in the same (waxbill) family—both males and females learn the song of their host fathers and will mate only with indigobirds that sing or respond to the right firefinch song, including the correct local inflection (see SONG). On the one hand, this improves the odds of cuckolding the right host species and maintaining this remarkable identity bond. However, if an oddball female indigobird places eggs in the "wrong" waxbill nest and the hosts do not reject them, the misplaced young are programmed to learn the ways of their foster parents, and the foundation is laid for the evolution of a new species of indigobird (see INTELLIGENCE).

BROOD PARASITISM AND BIRD CONSERVATION. A potential catch-22 in the brood parasite model is that theoretically if the parasite is too "successful," it could put its hosts—and therefore itself—out of business. Where small isolated populations of birds coexist with generalist parasites, local extinctions can occur. Sadly, there are already examples of the worst-case scenario, created largely by human agency. Cowbird populations and distribution have expanded greatly due to habitat alteration and an increase in food sources. This has enabled Brown-headed, Shiny, and other cowbirds to reach hosts from which they were once barred by ecological barriers. Were it not for muscular efforts to control Brown-headed Cowbirds in and around the Huron-Manistee National Forest in northern Michigan, the world's only population of Kirtland's Warbler might have vanished forever decades ago. And the endemic Yellow-shouldered Blackbird of Puerto Rico faces a similar threat from the rapidly increasing Shiny Cowbird, a recent arrival over much of the Caribbean and southern Florida.

Buzzard

In North America, this is a slang term for the New World vultures. Elsewhere in the English-speaking world it is the standard name for the relatively broad-winged, short-tailed raptors in the genus *Buteo* and similar forms. (In this Old World nomenclature, arising from the ancient traditions of falconry, our Red-tailed Hawk would have been properly called Red-tailed Buzzard and the term "hawk" would be reserved for the long-tailed raptors of the genus *Accipiter*.)

Turkey Vulture (*L*) and Common Buzzard

Caching

Caching
Woodpeckers, tits, nuthatches, and members of the crow family (including jays, magpies, and nutcrackers) habitually store food. The degree to which it is retrieved by the storers apparently varies and has been thoroughly studied in only a few species. Eurasian Nutcrackers are known to recover a majority of the nuts they bury in caches in the ground for sustenance during the winter and to feed young in the spring. And the North American Clark's Nutcracker stores food in communal caches and displays an uncanny ability to relocate these sites even when they become obscured by heavy snow cover.

Canada (Gray) Jays are also prodigious cachers. Patient ornithologists have counted more than 1,000 caches

Canada Jay

during a single 17-hour day. Fitted with exceptionally large salivary glands, the jays can make quantities of very sticky saliva with which they glue sundry morsels into tree crevices and other secure niches. The food items are stored within a prescribed radius from the sources, with the birds instinctively balancing the quality of the tidbit (i.e., its size or nutritional value), the distance it will have to travel to retrieve the cache, and the contingency of densely situated caches being robbed.

A key element of effective caching, of course, is retrieval: *Where the heck did I put that Cheese Doodle?* Retrieval success—essentially a function of memory—has been the subject of careful studies in a few species that demonstrated an extraordinary ability to remember not only where they stored some tidbit but what the tidbit was and how long ago it was cached. Clark's Nutcrackers, for example, can store more than 30,000 items in up to 2,500 locations over a season, in some cases traveling more than 15 miles, and are able to recover about two-thirds of them as much as 13 months later.

That many stored items go unrecovered is demonstrated, for example, by the sprouting of oak trees where jays were known to have buried acorns—adding seed dispersal to the benefits of avian caching.

Canopy Feeding

A practice peculiar to several species of herons in which the wings are spread forward to shade the area of water in which a feeding bird is standing, presumably affording some advantage in catching prey. It has been suggested that the canopy acts as a trap of sorts, with fish attracted to a suddenly available patch of shade. It also seems plausible that the gesture may act as a simple sun shade for herons feeding in the open, where the glare off the water obstructs a clear view of activity below the surface. It is not clear to what extent either or both of these possibilities explain this distinctive behavior.

Carson, Rachel (1907–1964)

That the general public, government agencies, and even corporate interests now understand the dangers of indiscriminate use of toxic chemicals in the environment can be attributed to a great extent to Rachel Carson and her consciousness-altering book *Silent Spring.* Though Carson focused her analysis on the effects of pesticides such as DDT, dieldrin, and heptachlor on birds, she made it clear that the outcomes were likely to be equally dire for human beings and for the Earth's biosphere as a whole.

The powerful impact of *Silent Spring,* which began even before it arrived in bookstores, resulted from the author's combination of scientific knowledge and a writing style that was both lyrical and accessible. For Carson, it was the writing that came first; she composed her first works at age 8 and was published by 10 (in *St. Nicholas Magazine*). And she began her college career as an English major before switching to biology.

Inspired by her reading as a teenager of Melville, Conrad, and R. L. Stevenson and her growing fascination with the ocean, she became an aquatic ecologist, earning her Masters with a thesis on the embryonic development of the excretory system of fishes. And her first full-length works, eventually to be known as the Sea Trilogy, celebrated the wonders of the marine environment. The first of these, *Under the Sea Wind* (1941), was critically acclaimed but did not sell well initially. But her next book, *The Sea Around Us* (1951), was serialized in the New Yorker, abridged in the Readers' Digest, won the 1952 National Book Award for non-fiction, and remained on the New York Times best seller list for 86 weeks. This allowed her to quit her job as chief editor  of publications at the U.S. Fish and Wildlife Service and concentrate full time on her writing. In 1955 she published the *Edge of the Sea*, which was also well received.

Carson became interested in the effects of pesticides on the environment as early as 1945 but could not capture the interest of publishers in the subject in the postwar era, when an emerging chemical industry seemed to promise the miraculous eradication of life- and crop-threatening insect pests. But by the late 1950s—due to programs such as the federal Gypsy Moth Eradication Program, which employed blanket aerial spraying of pesticides, and to disturbing events such as the "cranberry scare," in which herbicides were linked to cancer—she was able to publish articles documenting bird deaths attributable to pesticides, and in 1962, she published *Silent Spring*.

The effect was explosive—in both directions. Carson's well-documented and eloquently described thesis that pesticides were actually "biocides" potentially

affecting the well-being of nontarget organisms, including people, and that their residues accumulated in the
environment captured the public's imagination immediately and was soon taken up by conservation organizations, government commissions, the president's Scientific Advisory Committee, even the Supreme Court.
From industry, the outcry was equally vehement, though
often less civil. Her scientific credentials were questioned
because she was a marine biologist, not a biochemist.
One industry chemist thundered that if we followed
Carson's lead we would "return to the Dark Ages and
the insects and diseases and vermin would once again
inherit the earth." And another called her a "fanatic de-
fender of the cult of the balance of nature." Ezra Taft
Benson, a former U.S. Secretary of Agriculture, opined
that because she was unmarried, despite being physically
attractive, she was "probably a communist."

Given the widespread support of the public and
much of the mainstream scientific community, Carson's arguments prevailed and can be seen as inspir-
ing the founding of the Environmental Defense Fund
in 1967 and the establishment of the Environmental
Protection Agency in 1970. She is also credited with
influencing the rise of eco-feminism and empowering
many women scientists.

Carson's legacy contains many national and international honors, including the Presidential Medal of Freedom awarded by Jimmy Carter in 1980. Sadly, this and
most of the other tributes were awarded posthumously.
During her work on *Silent Spring*, she developed breast
cancer and died of complications from the disease on
April 14, 1964.

Chicken Hawk

Pejorative colloquial name used in the broadest sense to refer to any medium-sized to large hawk, any of which is presumed by the ignorant to ravage poultry yards habitually. The Northern Goshawk preys on grouse among other medium-sized birds, but it is a forest species unlikely to haunt barnyards. Smaller accipiters such as the Eurasian Sparrowhawk and Cooper's Hawk are more wide-ranging, know an easy kill when they see one, and might carry off a chick on occasion, but would hesitate to tackle an adult hen. In summary, the term has little basis in fact and has been one of many pretexts, especially pre-bird conservation laws, to kill raptors.

Climate Change

There is no longer any credible doubt that the Earth's changing climate and its living inhabitants are now on a collision course. Even so, there are many who refuse to acknowledge this, either from self-interest or because of understandable denial, based on a lack of knowledge, a distrust of science, a willingness to listen to those who say it is all a hoax, and perhaps an underlying anxiety that it is all true.

Phenomena such as shrinking glaciers, rising sea levels, ocean acidification, temperature averages, and weather anomalies don't raise alarms for many people because they tend to happen gradually and are comfortably dismissed as natural fluctuation: "The climate has always changed!"

Just as the warning of a "silent spring" by Rachel Carson galvanized the public to demand the banning of DDT in 1972, it may be that birds may once again provide the most visible evidence that a new threat has arrived—and that it is not just birds that will suffer from its effects. For example:

— Shorebirds that depend on coastal beaches are already losing real estate for nesting and roosting due to sea level rise, and higher tides are flooding Saltmarsh Sparrows out of their nearby (and only) breeding habitat. Tern colonies on barrier islands, like human inhabitants of low-lying archipelagos, are beginning to sink under water.

— The warming and acidification of the ocean causes a loss of biological productivity due to a reduction in dissolved oxygen. We are already seeing local shifts and reductions in populations of sea ducks, alcids, and pelagic species such as shearwaters and storm-petrels due to a reduction in the crustaceans, mollusks, and fish on which they feed and changes in the life cycles of the planktonic organisms that support the marine ecosystem.

— It is now well documented that plants in the Northern Hemisphere are blooming earlier due to earlier and warmer springs. Over eons, long-distance migratory birds have timed their arrivals at key stops as they move northward to their nesting grounds, to coincide with the flowering of prominent tree species. These attract myriads of insects that the birds need as fuel to continue their migration. If the birds, which cannot foresee changes of the flowering cycle from their distant wintering grounds, arrive too late to catch the peak bloom and pollination period, they may lack the fitness to rear young successfully or fail to nest altogether.

— The composition and microclimate of our forests are also changing with rising temperatures. Birds are extremely sensitive to changes in the structure of their habitats and the associated web of other organisms on which they depend. For example, as more humid forests in North America become drier they may cease to support a thick growth of understory shrubs that Wood Thrushes and other birds of the forest floor require for nesting cover.

In view of the many abundantly documented (and largely preventable) threats to birdlife—and human life—that we have visited upon the planet (see APOCALYPSE), and given that predictions for the effects of climate change are becoming more dire, not less, do we really want "let's wait and see" to be our motto going forward?

Cloaca

In birds, reptiles, amphibians, and many fish, the terminal enlargement of the digestive tract, through which solid wastes, urine, and the products of the reproductive system all pass prior to excretion, egg-laying, or copulation. In mammals, of course, the digestive passage is separate from the urinary and genital passages. The word derives from a Latin verb meaning "to cleanse," and the non-zoological meaning in English is sewer or toilet.

See also POOP, ETC. and SEX.

Colloquial Names of Birds

People the world over have given names to conspicuous birds or those that have some significant impact on their daily lives. Cultures that use birds extensively for food or adornment tend to have a high degree of recognition of their region's birds and invent names to differentiate them. These colloquial or local names almost always refer to some distinctive plumage, voice, or behavioral characteristic. In this regard, in fact, they are often more relevant and almost invariably more imaginative and pleasing than standardized vernacular (common) names (see below).

Of course, a majority of standard English names for birds derive from colloquial stock, especially at the generic level, and are sometimes of ancient lineage. "Finch," for example, can be traced back at least 3,000 years to a similar word that echoes (quite recognizably) the sharp call of the Eurasian Chaffinch; other, seemingly distinctive names, such as "Merlin" can be traced back only so far before disappearing in a tangle of linguistic roots. The American Robin would doubtless have been called Red-breasted Thrush or the like had it not reminded British colonists of the Robin Redbreast (European Robin) of their homeland. "Anhinga" and "caracara" come to us virtually unchanged from classical Tupi, a now-extinct language of natives of the Amazon Basin. And a significant number of "official" common names have survived intact from their folk origins, for example, Wheatear (a corruption of "white arse," referring to their distinctive rump and tail pattern), and Bobolink (originally "Bob Lincoln," echoing its bubbly song). As in this last example, it was once common in Britain to personify common bird species, hence Maggie the Pie which became Magpie and Jack the Daw, shortened to Jackdaw.

Despite the fact that we have now constructed a reasonably standardized English nomenclature for our native (and other) avifauna(s), it should be remembered (probably with considerable gratitude) that the great majority of people have paid not the slightest attention. A few of us insist that this pointy-winged bird, flying erratically over fields and cities at dusk uttering a nasal call that sounds like *peent!* is a "Common Nighthawk." But those who know it (in some ways at least) as well as or better than most ornithologists do, know full well that it is a "bullbat," which, after all, is shorter and hardly less accurate than the "official" name.

The main objection raised against colloquial names, of course, is that they are confusing. One person's Black-bellied Plover is another's "gump" or "chucklehead" and still another's "too-lee-huk." But it is on scientific names that we depend for nomenclatural consistency, and the terrible ambiguity exemplified above can usually be relieved with a little conversation. Furthermore, it is absurd to claim that standard vernacular names are especially apt. Golden and Blacksmith Plovers are just as "black-bellied" as *Pluvialis squatarola* yet, alas, no one had the imagination to call the Black-bellied Plover instead "Silver Plover." The smorgasbord of regional flavors and the local humor contained in North American bird slang easily outweigh any fancied need to cleave rigidly to an ornithological Esperanto.

It is gladdening to think that someone out there may still know the Ruddy Duck as a biddy, blatherskite, butterball, blackjack, hobbler, broadbill, bluebill, daub duck, dipper, dapper, dopper, bullneck, bumblebee buzzer, butter bowl, chunk duck, deaf duck, dinky, dip-tail diver, goddamn, goose teal, greaser, broadbill dipper, creek coot, pond coot, dumb bird, goose wigeon, stiff-tailed wigeon coot, hardhead, toughhead, steelhead, sleepyhead, hardheaded broadbill, booby, murre, pintail, hickoryhead, leatherback, leather breeches, lightwood knot, little soldier, muskrat chick, noddy paddy, paddywhack, quill-tail coot, rook, spoon-bill, gray teal, bumblebee coot, saltwater teal, shanty duck, shot pouch, spiketail, spatter, spoon-billed butterball, stiffy, stub-and-twist, water partridge, wiretail, and who knows what else. This is a mere 60-odd names; the naturalist John K. Terres has noted that there are at least 132 "common" names for the Northern Flicker.

Needless to say, it is impossible to give more than a sample of such names here. The names below were selected by the author for their historical, anthropological, ornithological, or etymological interest—or because they made him laugh.

Arsefoot: Refers to the loons and grebes, all of which have their legs situated at their extreme tail end, a useful adaptation for swimming and diving, which, however, makes moving on land awkward at best.

Burgomaster: Seamen's name for the Glaucous Gull, which tends to have a well-filled figure and a proprietary bearing. The name derives from the Germanic term for a chief magistrate, for example, a mayor.

Butcherbird: Northern and Loggerhead Shrikes, both of which kill large insects, small mammals, and birds and then hang the "meat" in the crotch of a branch or impale it on a thorn.

Callithumpian Duck: Long-tailed Duck (until recently Oldsquaw); a callithumpian band is an amateur musical group that characteristically produces an odd assortment of notes at random; the same may be said of flocks of Long-tailed Ducks.

Butcherbird

Doughbird (or doebird): Though Thomas Nuttall and many shorebird shooters came to regard this name as generic for many of the larger, longer-billed sandpipers, it originally referred specifically to the Eskimo Curlew, "for it was so fat when it reached us in the fall that its breast would often burst open when it fell to the ground, and the thick layer of fat was so soft that it felt like a ball of dough." (Bent, 1929)

Erne: An old Anglo-Saxon name for the White-tailed Eagle, perpetuated in modern usage in the Scandinavian words for eagle (ørn or örn); well known to workers of crossword puzzles.

Jiddyhawk: Seamen's name for the species of jaegers; this is a laundered version of a name that refers to jaegers' alleged habit of eating excrement expelled by terns and other birds that they harass. The Latin genus and family names of the jaegers (*Stercorarius,* Stercorariidae) also mean "eaters of excrement."

Mother Carey's Chickens: Seamen's name for the storm-petrels, particularly European, Wilson's, and Leach's.

Speculation that this widespread nickname is associated with the Virgin Mary (i.e., *mater cara* or *mater caritas*) seems at odds with the usual superstitions about "carey chicks," which describe them as tormented souls of lost sailors or cruel ship's officers or even avian demons keeping watch over the drowned. It is more likely a so-called noa name, that is, a name used to avoid calling an evil spirit by its right name and thus attracting it.

Peabody Bird: White-throated Sparrow, one song of which is often verbalized by residents of the United States as "Old Sam Peabody-Peabody-Peabody"; Canadian ears hear "Oh, sweet Canada, Canada, Canada."

Preacher: Two-way pejorative name for the Red-eyed Vireo, which is noted for its monotonous, endlessly repeated song.

Shitepoke: Laundered version of a name applied to several species of herons, referring to their habit of defecating conspicuously and often copiously when flushed.

Smutty-nosed Coot: Hunters of North Atlantic waterfowl call all of the scoters "coots." This variation refers to the fatty orange knob at the base of the bill of the male Black Scoter.

Tickle-arse: Seamen's name for the Black-legged Kittiwake; the name is a contraction of "tickle-your-arse-with-a-feather" and is inspired by the giggle-like calls of this northern gull; "kittiwake" and "tickle-arse" are also onomatopoeic for this species' titter.

Timberdoodle: One of many folk names for the American Woodcock, an upland shorebird that nests in woodlands but performs dramatic aerial displays over adjacent open areas. Other names include bogsucker, hokumpoke, twitter pate, and bumblebee chicken.

Wobble: Seamen's name for the Great Auk, perhaps describing its penguin-like gait? Not to be confused with "wobbla," the collective name in New England for the members of the family Parulidae, for example, "a wobbla wave at Mount Awbun (Auburn) Cemetery."

Convergence (Convergent evolution)

The development of similar traits in unrelated groups of organisms due to adaptation to similar living conditions. In a broad context, the development of the forearm into a wing in both birds and bats is an example. In birds, a number of uncanny look-alikes from different families have evolved in this manner. Two striking examples are the superficial resemblance between some members of the Northern Hemisphere auk family (Alcidae) and the phylogenetically distant penguins and diving-petrels of southern oceans. Other examples include the New World hummingbirds and Old World sunbirds and the nuthatches (Northern Hemisphere), Nuthatch Vanga (Madagascar), and sittellas (Australasia), all unrelated.

Coues, Elliott (1842–1899)

One of the foremost—and easily the most interesting—of the American ornithologists of the late nineteenth century. Coues was born in New Hampshire, graduated from the college and medical school of what is now Georgetown University in Washington, D.C., and almost immediately began a writing career that eventually included almost 1,000 works, several of them major volumes. He was the most eminent of the army surgeons who collected specimens for Spencer BAIRD while on duty in the western states and became secretary and naturalist of the U.S.-Canadian Border Commission, out of which experience came two of his own ornithological works, *Birds of the Northwest* and *Birds of the Colorado Valley*. He also published extensively on western mammals. As secretary and naturalist for the Geological and Geographical Survey of the Territories, Coues added papers on the history of the exploration of the American West to his extensive bibliography. He was a founder of the American Ornithologists' Union (now renamed the American Ornithological Society), and his 1882 *Check-list of North American Birds* was the basis for the AOU publication of the same name, the first edition of which appeared shortly thereafter.

What set Coues apart from the other brilliant naturalists of his time was his personality, which has been described as "electrifying." He was physically handsome and had a highly developed sense of humor, boundless energy, and a greater than average taste for the eccentric. The most notable example of this last characteristic was his passionate conversion late in his life to the cult of the infamous "spiritualist" charlatan Madame Blavatsky, whose séances were also attended by such worthies as Thomas Edison, John Ruskin, and William James. Coues took to spiritualism with the same zest he brought to his other interests but was eventually banished from the movement for publishing the details of a Blavatskian hoax. Coues has remained unchallenged as the best writer among American ornithologists, and much of what was in his personality is evident in the eloquent, witty, and opinionated introductory chapters to his *Key to North American Birds*. Besides being a landmark in American ornithology, this volume contains a fine evocation of what being a nineteenth-century naturalist was like, from how to clean your collecting gun ("elbow grease"), to the wisdom (none) of taking "stimulation" when afield, to the perils of skinning a putrid bird (festering sores), to the excesses of the overzealous "splitters" who dominated taxonomic theory in Coues' time. Until the 1980s, the Greater Pewee (*Contopus pertinax*) was known as Coues's Flycatcher. Given his distinguished ornithological career, the loss of his eponym seems a pity.

Crepuscular
In general usage, refers to twilight. But in the context of animal behavior, the term means active in low levels of light, especially at dusk but also before dawn. No

Woodcock courtship display

bird species is exclusively crepuscular, but some species of owls, swifts, and nightjars are especially active or conspicuous in the twilight hours just after sunset and around first light. Some shorebirds (e.g., woodcock and snipe species) and songbirds (e.g., Henslow's Sparrow) are especially active at these hours, especially in the context of courtship. The skimmers, which feed best in calm water, typically take advantage of the low wind levels characteristic of dawn and dusk and are therefore largely crepuscular in their feeding habits. Many mammals and insects, and some reptiles are also crepuscular.

Display

In the broadest sense, any innate, stylized visual signal made by a bird, the function of which is to trigger or "release" appropriate behavior in the intended object of the signal. The most conspicuous examples of avian display are complex, prolonged rituals in a sexual or defensive context. These frequently involve the actual "display" of some prominent plumage or other physical characteristic, but the gestures involved often give no hint of the intended outcome—at least to a human witness. All birds engage in some form of display, though there is much variation in both size of repertoire and intricacy of ritual.

COURTSHIP (SEXUAL) DISPLAYS. When courtship displays are performed by males alone, they serve not only to attract the attention of unmated females, but also as "no trespassing" signs to other males. The displays of most ducks, shorebirds, hummingbirds, and passerines are dominated by the males. While he engages in some form of plumage display, accompanied

Male frigatebird inflating red throat sac to attract females

by sound effects and/or movements, the female typically remains composed, sometimes appearing completely indifferent, even impatient, and may express her reaction with "displacement" activities such as feeding or preening. Or she may respond with ritualized gestures of submission alternating with solicitation. For one elaborate form of this kind of display, see LEK.

Many waterbird species in which the sexes look alike—for example, loons, grebes, tubenoses, pelican relatives, gulls, and terns—perform courtship displays in which both members of a pair are equally active. Some large grebes, for example, engage in tandem "races," and gannets face off in a mirror-image "greeting" ceremony.

The often bizarre and elaborate antics of birds engaged in sexual display are among the most fascinating and entertaining aspects of birdlife. Virtually every avian attribute and external anatomical feature—plumage, bill, legs, eyes, voice, flight—is active in one ritual or another, and not a few adaptations apparently function solely in display. Egrets, for example, acquire exquisite, erectile head and back plumes (historically called "aigrettes"); many alcids develop decorative facial tufts and colorful bill sheaths; and the colors of the so-called soft parts (bill, legs, fleshy eye rings, and facial skin) intensify greatly, though often very briefly, in a wide range of species during courtship.

FLIGHT DISPLAYS are a class of mating/pair bonding ritual that takes many forms, for example:

— *Volplaning.* In this display, a bird or pair of birds glide down from a height with wings held outstretched and motionless. Red-billed Tropicbird pairs volplane one above the other with the upper bird's wings held down and the lower bird's up so that they almost touch; courting terns often volplane (sometimes in groups); and some sandpipers volplane down from song flights. Perhaps the most readily seen volplaning displays are those of many common species of pigeons and doves over their long breeding seasons.

— *Butterflying.* Many plovers and sandpipers incorporate a very distinctive slow, shallow wingbeat in their flight displays, as do male Vermilion Flycatchers and American Goldfinches.

— *Hovering.* This replaces a tall singing perch for many open country species, for example, larks, pipits, longspurs, and many tundra-breeding shorebirds. The singer may hover hundreds of feet in the air while delivering a full-throated aria.

— *Plummeting.* Woodcock and snipe descend from the top of their aerial displays in erratic swoops, making characteristic mechanical sounds (see SONG) as they do so. Common Nighthawks plummet and "pull-out" abruptly after gaining speed, making a distinctive noise with their wings as they arc upward.

— *Sky Dances.* Male hummingbirds describe patterns in the air characteristic of their species in front of their mates, calling and in a few cases also making feather noises.

— *Aerobatics.* Perhaps the most dramatic of aerial displays are the stunt maneuvers performed by most species of hawks and eagles. A familiar variation is a series of continuous undulations. Some species fold their wings and plummet from as high as 1,000 feet before pulling out sharply and rising to repeat the dive. Vying for most remarkable stunt is the "tumbling" of some eagle pairs. In the North American Bald Eagle, the male flies above the female. She then turns on her back and presents her talons, which the male grasps in his. Then they fall down through space in a tumbling roll, releasing each other in good time. Many species of hawk also lock talons during courtship displays, falling earthward in helicopter-fashion, like some enormous maple samara.

DANCING. Pairs and sometimes small groups of albatrosses, cranes, some pelican relatives, and many gulls and shorebirds engage in more or less conspicuous, formalized, earthbound displays. These involve a great diversity of actions, including "bill fencing," gaping (to show the bright color of the mouth lining), wing posturing, "curtseying," pointing, stretching, and presentations of food or nesting material. Owing to their size and the exaggerated and intricate choreography of many of their movements, the albatrosses and cranes are especially renowned as "dancers."

WATER DISPLAYS. The courtship displays of loons, grebes, and waterfowl are essentially aquatic dances. Loons and grebes with legs placed near the end of their bodies are capable of rising nearly erect from the surface and paddling furiously across their breeding lake in this vertical posture; pairs do this in tandem as one act in the courtship ritual. The goldeneyes execute a distinctive combination of head and neck motions involving "inflating" the head feathers until the head is nearly spherical, stretching the neck forward, and then tossing the head violently over the back ("head throwing"); in one version, this is accompanied by splashing, which exposes the bright orange legs and feet; other ducks have their own balletic moves.

THREAT DISPLAYS. Faced with territorial intruders of their own or other species, most if not all birds assume characteristic defensive postures, whether to intimidate rivals or protect themselves and their young. The tendency to erect body feathers with the effect of making the potential victim seem larger or otherwise more impressive is a common ploy. Blue-footed Boobies, for example, create an imposing presence by raising the head and neck feathers, making them bristle like porcupine quills. Nestling hawks and owls and cornered adults will often spread their wings and display their talons as well as raising their feathers as a means of intimidating attackers.

Possibly the most impressive achievement in the evolution of threat performances is the so-called snake display enacted by species of titmice. When their hole nest is invaded, the sitting bird gapes wide, hissing and swaying in a serpentlike manner and finally "strikes" upward, simultaneously hitting the nest wall with its wings.

Of course, when threat displays fail, a counterattack is sometimes mobilized—as anyone who has ever wandered into a tern colony or ventured too near a skua's or goshawk's nest during the breeding season can attest.

DISTRACTION DISPLAYS. This type of display by nesting adults diverts the attention of predators from eggs or young. It is most highly evolved among open-country ground nesters, but it has been observed in some form in most bird families. The basic effect, performed with greater or lesser verisimilitude by different species, is of an injured or ill bird, flopping helplessly on the ground (away from the nest) while crying out in faux anguish. This "broken wing act" performed by many sandpipers and plovers is a remarkably convincing bit of avian pathos.

Diving

A significant proportion of the world's living birds habitually dive in water. This includes the penguins, loons, grebes, some tubenoses, some pelicans, tropicbirds, boobies and gannets, cormorants, anhingas, coots, sungrebes, terns, alcids, some raptors, about 60% of duck species, some kingfishers—and one songbird family (see Dippers, below). And other swimming birds such as shearwaters and gulls will dive occasionally for various reasons explained below. The behavior is mainly a means of obtaining food but can also be effective in eluding predators. Some characteristics of and statistics about diving birds are as follows:

FOOT POWER. When progressing normally at the surface, diving birds paddle with alternating strokes. But birds that use foot power while diving push with both feet at once.

WING POWER. Wings modified for diving are surprisingly common among birds, since most adaptations of these limbs are related to flight. The most extreme example is the penguins, whose forelimbs have become flippers—useless for flying in the air but perfectly adapted for "flying" underwater. Of all birds, penguins are by far the most at home in the water. They are as agile as any marine mammal (or more so) and can swim underwater faster than many birds can fly—Gentoo penguins at up to 22 miles per hour. The alcids—petite penguin lookalikes of the Northern Hemisphere—have wings that are also strongly adapted for use underwater. Though fully feathered and capable of flight in living alcid species, their forelimbs are notably narrow and flipperlike when extended. Some shearwaters and petrels also use their rather narrow wings to "fly" under water in pursuit of prey in ocean waters; some of these "tubenose" species regularly dive 100–200 feet deep in this manner.

LIGHT HEAVYWEIGHTS. Birds that habitually dive should ideally both be buoyant and have a high specific gravity. This seemingly contradictory condition is achieved in most divers by a combination of anatomical specialization and behavior. Birds' body cavities are lined with a system of air sacs that fill the spaces between the other organs and can be inflated and deflated as the lungs are by inhaling and exhaling. These air sacs even penetrate many bones so that even a bird's skeleton is partially air filled. The buoyancy thus achieved is of course an advantage in flight, and, when combined with the air-trapping capacity of PLUMAGE, also makes birds natural floaters.

When it comes to diving, however, buoyancy becomes a liability, working against a bird's efforts to penetrate water, which becomes ever denser with increasing depth. Species that plunge into the water from the air depend partly on gravity to counteract their natural "floatability," and those that dive from the surface and pursue prey once submerged push themselves under using their legs and feet—and in some cases their wings (see above).

More subtle is the ability of many waterbirds to compress their plumage, squeezing trapped air out and raising their specific gravity. It is this trick that allows the grebes to sink like a submarine apparently without exerting any effort. The plumage of cormorants and anhingas is relatively porous and fills with water readily; this makes for efficient diving but also explains the amount of time these birds spend drying out with their wings spread. In preparing for a dive, birds also expel air from their lungs and air sacs, further reducing buoyancy.

OUCH? Anyone who has watched a gannet plummet head first into the sea from high in the air is likely to wonder how the impact can be sustained time after time without resulting in a broken neck or at least a bad headache. In its practical way, evolution has given gannets and boobies unusually powerful neck muscles, a thickened and strongly supported skull, and air sacs lying under the skin of the head, which act as a kind of crash helmet. The eyes of diving birds are also modified to enable them to see prey underwater (see VISION).

And the nostril openings of pelicans, cormorants, and gannets and boobies are either closed at birth or become so by adulthood, and breathing is done through openings at the corners of the gape, another adaption to underwater life.

HOW DEEP AND HOW LONG? The farther down a bird dives, the harder and slower the going, and therefore there is some correlation between the depth of a dive and its duration. However, long dives are not necessarily deep dives: one Western Grebe submerged for 63 seconds in 5½ feet of water.

The majority of diving birds are not deep or long divers; most species rarely exceed 10 feet or stay under more than 10–20 seconds. Even the deep divers, such as the loons, grebes, sea ducks, shearwaters, and alcids rarely approach their full diving potential. The record-holding Common Loon (see below) usually dives to a depth not exceeding 35 feet and takes less than a minute (sometime much less) to resurface. Also, birds under duress (e.g., chased or wounded) will typically stay submerged much longer than normal.

RECORD DIVES. The Emperor Penguin can reach at least 1,754 feet (534 meters) below the surface and stay under for at least 16 minutes, currently the world record. The Common Loon and the Long-tailed Duck are tied at dives of 180–225 feet and perhaps slightly deeper, both having been trapped in fishing nets set at these depths. Though their average dives are much shorter (see above), these birds are apparently little taxed by dives lasting 3 minutes, and the loon has survived forcible submersion for 15 minutes. These follow only the Emperor Penguin as world records.

DIPPERS (5 species in genus *Cinclus*) are the only songbirds that habitually swim and are quintessentially birds of fast-moving streams and rivers. They superficially resemble large (6-inch), stocky wrens with their compact form, short wings, frequently "cocked-up" tail, and bubbly song. Closer inspection reveals a laterally flattened, slightly hooked bill, an unusually thick feather coat and underlying down, and somewhat pointed wings and proportionately large, strong feet—all adaptations to their unique lifestyle. This involves plunging into the icy water (they can dive to a depth of at least 20 feet) and then walking on the bottom of the torrent, picking aquatic insect larvae, snails, small fish, and the like from among the submerged rocky crannies. Periodically, they rise to the surface, float buoyantly to a nearby rock and preen their feathers. Dippers occur mainly in mountainous areas throughout much of Eurasia, western North America, and the Neotropics.

Drunkenness

Birds occasionally become intoxicated from unintentionally ingesting fermented fruit or nectar, and instances of fatalities resulting from lack of control have been recorded. Waxwings, some thrushes, and other frugivores that habitually depend on "old" fruits of mountain ash, crab apple, or similar species in winter are sometimes killed outright by the alcoholic toxins contained in the fruit or fly into windows or other objects after becoming tipsy. Similarly, *Agave* blossoms, which are very popular with a wide variety of nectivorous birds in warm climates, ferment rapidly when filled with rain water, which then "boils" in the hot sun. In at least one case, a highway near such an organic cocktail lounge was littered with squashed corpses of avian tipplers that apparently were rendered less sensitive to the danger of passing traffic and/or lost some motor coordination.

Rufous-and-white Wrens

Duetting

In birds describes a pair of birds (or members of an extended family) singing together—either (1) the same song in unison or (2) in closely synchronized alternating phrases, known as "antiphonal" song. In the "classic" form of the latter, the male and female of a pair each sings a characteristic *different* phrase with the phrases so closely spaced that the effect is of a single song. Duetting is mainly a tropical phenomenon and has been recorded in more than 200 species in 44 families. Most duetters are monogamous species that defend breeding territories year-round. Presumably, duets help solidify the pair bond as well as maintaining territory; they also allow birds to locate each other easily in dense habitats.

See also SONG.

Dummy Nest

A phenomenon, especially characteristic of wren species, in which the male, arriving on territory ahead of the female, constructs a series of shells, that is, external nest structures without linings. When the female arrives, she is taken on a songful tour of inspection, following which she selects one of the nests and lines it or makes other substantial improvements in preparation for egg laying. After the breeding season, the dummy nests (a.k.a. "cock nests") may be used by young and adult birds as "dormitories."

Dump Nest

A phenomenon characteristic of some ducks and gallinaceous species and the South American rheas in which several females lay eggs in a single nest that is ultimately abandoned. In at least some cases, dump layers are birds with no nests of their own and also tend to be of species that sometimes practice BROOD PARASITISM. In a nest in which only a few extra eggs are dumped, the owner may incubate the clutch, even if it includes eggs of a different species. However, when the egg pile reaches intimidating proportions—one pochard nest was found to contain 87 eggs—the hen abandons it.

Nest dumping differs from communal practices of ostriches, tinamous, anis, and other birds that lay eggs in a single nest and share incubation duties.

Dusting (Dust-bathing)

Some bird species actively and regularly fill their plumage with fine, dry soil or sand and then remove it by shaking and preening in a ritual known as "dusting" or dust-bathing. The practice seems to be most prevalent among birds of open country, such as larks, gallinaceous birds, and some (emberizine) buntings, but it has also been recorded in some hawks, doves, owls, nightjars, wrens, Wrentit, Common Grackle, and House Sparrow. Some "dusters," for example, gallinaceous birds, apparently never bathe in water, and larks only bathe in water by exposing themselves to the rain. But other birds bathe in both water and dust. According to Simmons (1985), dusting and ANTING are mutually exclusive as far as is known, except for one species of Australian songbird, the White-winged Chough.

While some dusting gestures—fluffing of feathers, shaking of bill in dust, flicking of wings—are similar to water-bathing gestures, the process is not truly analogous either in performance or in function. The ritual varies from species to species. Songbirds typically make a hollow in the dust by scratching with their feet and rotating their bodies, and the dust is applied by a combination of direct wallowing and the water-bathing gestures just described. Gallinaceous birds and others use bill and feet both to gather dirt around them and to "drive" it into or throw it onto their plumage. After a thorough dusting, birds shake and flap vigorously to rid themselves of the dirt. House Sparrows typically dust in small groups, and groups of turkeys (hens with poults) have been seen bathing together in dusty roads, but for the majority of species it is a solitary activity.

The purpose of all this is far from clear. Present evidence suggests that dusting helps keep plumage fluffy (hence air-retentive for good insulation) by removing excess moisture and preen oil. It is also a means of flushing out ECTOPARASITES such as bird lice; it may also reduce the abundance of keratin-consuming bacteria.

For related behavior, see PREENING; SUNNING; ANTING.

Ectoparasite

An organism that lives *on the surface of the body* of another organism, typically to the detriment of the host.

Birds are host to an astonishingly populous and varied fauna of invertebrate ectoparasites. Thousands of species and subspecies of flies, fleas, lice, ticks, and mites have been identified as living to some degree and in various ways on the bodies of birds. A single small songbird may have an unwelcome guest list numbering dozens of individuals of several species, and in the breeding season, its nest is likely to contain several times more. This will confirm for some the prejudice that birds are "dirty" and dangerous, yet most birds are able to control these pests to a tolerable level (at least from the bird's viewpoint), and transmission of either parasites or parasite-borne diseases from birds to people is probably rare (however, see below).

A few of the more frequent parasite lifestyles are worth noting in a general way. Some bird parasites, notably the bird lice and feather mites, live out their entire life cycles on the bodies of their hosts. Others feed on the host only at intervals or during one stage of their development and live in the nest, in the ground, on another host, on flowers, or elsewhere during other stages. Consequently, how birds become infested also varies.

Parasites that live exclusively on bird hosts are passed directly from bird to bird by direct contact. This occurs, of course, between parent birds and their young and also between raptors and their avian prey (hawks and eagles, vultures, and owls often accommodate an unusually rich variety of pests). Bird lice occasionally hitch rides from host to host on the bodies of larger and more mobile louse flies (family Hippoboscidae). Parasites that live part of their lives "ex-avis" contact birds in diverse manners. The eggs of some flies are laid in birds' nests and progress through larval and pupal stages during a single breeding season, parasitizing nestlings as larvae and quitting the nest as adults; other flies overwinter in the nest as larvae. Many parasites live in the nest but visit their hosts only at night. And most ticks fall off their host after a single blood meal and must find a new host for the next feeding.

A wide range of tolerances is exhibited by invertebrates that parasitize birds. Some adult chiggers (see below) are equally content to feed on the blood of birds, mammals (including people), or reptiles. Conversely, bird lice and feather mites are restricted not only to birds but also usually to certain types of birds (see below). Some are habitat-specific and may be attracted, for example, only to marsh-haunting birds or birds of arid scrub; the nest fly (family Neothopilidae) prefers nests made partly of mud such as those of many thrushes.

Families and species of bird lice tend to parasitize particular orders and families of birds exclusively and to parallel the phylogenetic relationships of their hosts. Thus, gull lice may be different from tern lice but are likely to be more closely related to each other than either is to duck lice. The remarkable parallel has proved so consistent that comparison of lice has been used to argue taxonomic relationships.

Food preferences of ectoparasites also vary. Fleas, flies, ticks, chiggers, and some lice suck blood, but feather mites and certain lice attack feathers (both the webs and the core of the shaft), and skin debris, body grease, and other bodily fluids are specialties of a number of species.

It may seem that a bird's chances of surviving the onslaughts of this army of tiny attackers are slim, and there is no question that ectoparasites can cause debilitation and death. Some are vectors of diseases fatal to birds; the toxic saliva of some ticks can cause death; an unusually heavy burden of any type of blood-sucking parasite can readily kill nestlings and severely weaken adult birds, making them more vulnerable to larger predators and disease; and the feather eaters and itch mites, which cause birds to scratch until raw, sometimes cause extensive feather loss, resulting in impaired flying ability and exposure.

Still, birds are not totally defenseless. Preening appears to be crucial, and birds unable to preen through some disability are more vulnerable to parasites. Practices such as DUSTING, smoke bathing, SUNNING, and ANTING may be, at least in part, methods of pest control. And, finally, among the array of tiny creatures that

inhabit birds and their nests are a number that feed
not on birds themselves but on bird parasites! One is
reminded of that venerable lyric summary of invertebrate biology:

> Big fleas have little fleas
> Upon their backs to bite 'em.
> Little fleas have smaller fleas,
> And so, *ad infinitum*.

Some of the most important ectoparasites are described below with brief descriptions of their appearance, habits, and importance. All are arthropods (phy-
lum Arthropoda). Flies, fleas, and bird lice are insects
(class Insecta); ticks and mites belong to the same class
as spiders (Arachnida).

Flies (order Diptera)

BOT FLIES (family Oestridae). Adults are large, hairy,
and somewhat beelike in appearance. The eggs are laid
on nestlings; the larva hatches and burrows into skin,
where it feeds until ready to pupate. They occur mainly
in the tropics.

LOUSE FLIES (family Hippoboscidae). These are the
most important fly parasites on birds, and a few species
(e.g., the wingless sheep ked) attack mammals. Adults
are usually smaller than a housefly, flat and "leathery"
looking, winged or wingless, and feed on their hosts'
blood. Wingless forms crawl through feathers and resemble crab lice. Adults remain on the host for life; the
larva hatches and matures within the body of the female
fly and pupates as soon as it is "born"; the pupa may
overwinter in the nest or in the ground. Hippoboscids

tend to be habitat- (i.e., species-) specific, but some are generalists and—logically enough—all species can be found on bird predators. There are about 150 bird-feeding species worldwide, and they have been recorded from at least 24 orders of birds. In birds that reuse nests year after year, hippoboscid pupae overwinter in the nest and re-infest the following generation. These flies can occur in relatively large numbers on relatively small birds, for example, 30–40 on species of swifts and swallows. They transmit trypanosomes and the blood parasite *Haemoproteus* but in general do not seem to greatly injure their hosts.

BIRD LICE OR CHEWING LICE (order Phthiraptera, formerly Mallophaga). Perhaps the most important bird ectoparasites, these are small, usually flattened, and wingless. There are more than 800 North American species in several families, and they are known from more than 500 bird species; a few prefer mammal hosts. Most birds are host to a number of species in different body "niches": head and neck lice are fat, rounded, and sluggish with clasping mouthparts, and back lice are slender and agile to elude preening; these do not suck blood but bite off bits of skin or eat feathers and the pith of the rachis. A few specialist species live in throat pouches of cormorants and pelicans. Bird lice spend their entire life cycle on the bird, attaching eggs to feathers and dispersing by contact between birds. Heavy infestation may coincide with debilitation of the host and the lice may die with the host. They are highly host-specific, and phylogeny parallels that of hosts; thus, they are useful as taxonomic characters. Prodigious "blooms" of lice sometimes occur on moribund

Philopterid bird louse

birds, apparently as an effect rather than a cause of the individual's condition.

TICKS AND MITES (subclass, formerly order, Acarina) include many important bird parasites in about a dozen families of this large order. Behavior patterns are very variable with regard to host specificity, as well as feeding and breeding habits, and these creatures may occur in nests in spectacular numbers.

FEATHER MITES (Analgesidae and other families) are important parasites which, of course, are restricted to birds. Like bird lice, they spend their entire life cycle on their host, laying eggs on feathers or within quills. They are often host-specific and habitat-specific on individual birds, different species preferring different types of feathers and even different parts of particular feathers. Heavy infestation may severely damage plumage.

BIRD PARASITES AFFECTING PEOPLE. The major issue here is the transmission of disease organisms to people via mosquitoes or ticks. Birds are known to host ticks carrying *Borrelia burgdorferi*, the Lyme disease bacterium that is transmitted to people by *Ixodes* ticks, and also known to serve as reservoirs, as are Deer Mice and White-tailed Deer. Birds also carry West Nile and equine encephalitis viruses. These diseases affect both mammals and birds, and the latter is most virulent in birds. When transmitted to people by the bite of a mosquito they can be fatal, especially to the most vulnerable individuals, such as children and the elderly.

Edibility (of birds, eggs, nests)

In the present era of Migratory Bird Acts, strict hunting seasons, and heightened "eco-consciousness," the notion of eating birds suggests to most people the grocery store much more strongly than the wilderness. Yet as late as 1882, one O. A. Taft, proprietor of a hotel and restaurant at Point Shirley, Boston Harbor (famous for its "game" menu), could bet an assembled company of friends that they could not name an edible North American bird that he could not produce instantly. It is reported that he had no takers. Lest the reader have too narrow a conception of what Mr. Taft considered edible, a menu from his establishment that has been preserved includes the following species (translated where possible by the present author): Owls "from the North," Chicken Grouse (Prairie-Chickens) "from Ill.," Dough Birds (Eskimo Curlew), Willet "from Jersey," Jack Curlew (Whimbrel), Rock Snipe (Purple Sandpiper), Golden-plover, Beetle-head (Black-bellied) Plover,

Redbreast Plover (Knot), "Seckel-bill" (Long-billed) Curlew, Chicken Plover (Ruddy Turnstone), Summer (Lesser) Yellowlegs, Winter (Greater) Yellowlegs, Reed Birds (Bobolinks) "from Del.," Brown Backs (Pectoral Sandpiper), Grassbirds (Baird's Sandpipers), and Peeps (small sandpipers). The pièce de résistance of this particular bill of fare is Hummingbirds Served in Walnut Shells! A good crowd easily—and apparently frequently—disposed of a thousand such "game birds" of an evening at Taft's. And though by the 1890s a few voices were being raised against the slaughter required to stock such sumptuous larders, one could still buy songbirds and shorebirds in market game stalls in many cities in the first decade of the twentieth century. Even today, the occasional (illegal) deep-dish robin pie or other such fare finds its way to the table in rural areas.

Plume hunting and "pure sport" have, of course, taken their toll of birds, but the most important single encouragement to man's inclination to kill birds is that they are very good to eat. Some are better than others, of course; many—consider vultures, for example—do not recommend themselves because of their habits. Yet it is not unreasonable to suppose that virtually every species of North American bird has been eaten at least once by someone.

TASTE. It is gospel among some hunters that how a bird tastes depends on where it lives and what it eats. By this rule, nearly all seabirds are despised by some as "fishy" and coots rejected as "muddy." Though there is some truth to this generalization—as most people who have sampled cormorant flesh will affirm—it is belied by certain ingenious methods of cleaning and preparing "strong" birds, which make them very tasty or at worst innocuous. The author has heard recipes for virtually every species of seabird in the North Atlantic from experienced gourmands of Newfoundland. Many of these involved immediate evisceration and "blooding" and often the use of marinades ("pickles") in which strips of breast meat are cured for weeks or longer before cooking.

Another recipe, this one for the notoriously unpalatable sea ducks called scoters (or "coots" as they are nicknamed along the coast of New England), was offered by R. W. Hatch in the December 1924 issue of *Field and Stream*:

> The easiest way is to place the coot in a pot to boil with a good flatiron or anvil. Let it boil long and merrily, and when you can stick a fork in the flat iron or the anvil, as the case may be, then the coot will be ready to eat. If that takes too much patience, take the goodly coot and nail it firmly to a hardwood board. Put the board in the sun for about a week. At the end of that time, carefully remove the coot from the board, throw away the coot, and cook the board.

Other New England coot fanciers allow that it is preferable to roast the coot with a brick in the cavity and when the latter becomes soft, it (the brick) is ready to eat.

In addition to insisting that birds are what they eat, common knowledge also holds that the bigger and/or older a bird, the tougher and "gamier" the meat. Here, again, there is both support and refutation from those who have dined on ancient swans and the like.

A more scientific note has been sounded by ornithologists, who have shown that there is some correspondence between cryptic coloration and palatability in birds (and eggs) and that some very brightly colored birds are notably distasteful, which their conspicuous plumage advertises to potential predators—as with certain butterflies.

GAME BIRDS. Many wild species in the duck, turkey, grouse, pheasant, rail, and dove families are still shot legally (or not) in parts of the developed world for their tasty flesh. Most wildfowl eaters agree that certain pochards, teal, and Brant deserve special praise among waterfowl, provided, of course, they have fed in the right places and are prepared properly.

SHOREBIRDS. The steep decline of shorebird populations by the turn of the nineteenth century is directly attributable to market gunning. The Eskimo Curlew or Doughbird, once of "sky-darkening" abundance, is now almost certainly extinct, a consequence of its superb flavor (second, according to many, only to the Passenger Pigeon). By all accounts most of the sandpipers and plovers make excellent eating. Woodcock and snipe are the only "shorebirds" that may now be hunted legally in North America.

SONGBIRDS. Larks, thrushes, buntings, and many other passerines have been traditional food in civilized Europe for centuries and were legally trapped and eaten in large numbers in many countries until very recently; the practice continues locally but is now prohibited by law in the European Union. They are usually skinned, roasted on small spits, and eaten bones and all. Many southern European householders set window traps for House Sparrows, which some claim are only marginally inferior to Ortolan Buntings, for instance. House Sparrows are not protected by law in North America.

EGGS. "Egging" at seabird colonies was once widespread; it was at least partially responsible for the demise of the Great Auk. Inuit and other indigenous North Americans are still allowed by law to harvest the eggs from seabird colonies.

Seabird eggs, like their parents, are widely thought to taste "fishy." This may be the case under certain circumstances, but in the experience of many, a fresh Herring Gull egg is superior in taste to a supermarket hen's egg.

Plover, gull, and quail eggs are familiar European delicacies—the latter, at least, being provided by domestically bred birds.

The British ornithologist Hugh Cott has said that eggs of larger colonial nesters taste better, in general, than those of smaller solitary nesters and that (as with adult birds) cryptically colored eggs tend to taste better than conspicuous ones. The latter (he says) are likely to be bitter.

NESTS. Birds' Nest Soup is an expensive Asian delicacy made from the nests of several species of Old World swifts in the genus *Aerodromus*, which breed in large colonies in caves (and, increasingly, in buildings created solely as nesting places for them—nests sell for very high prices in China). The nests are made of viscous, solidifying saliva secreted by the swifts for the purpose of attaching their nests to the vertical cave walls. The

Edible-nest Swiftlet cave

saliva itself is relatively tasteless and is improved in most soup recipes by the addition of vegetables and condiments. Swift spittle is considered an aphrodisiac by the Chinese.

Egg(s)

In the broadest sense, an egg is simply the female reproductive cell or ovum, whether originating in the body of a female butterfly or human. A bird's egg, however, consists of the ovum, including its attached food supply (yolk), surrounded by a mass of gelatinous "egg white" (albumen), the whole encased in a hard calcium shell. It is this eggshell that typically captures the attention of people interested in birds' eggs.

THE EGGSHELL is secreted from the walls of the shell gland—a portion of the oviduct, which in some respects is analogous to the human uterus. The suspended calcium carbonate salts that are secreted quickly harden into interlocking crystals (calcite), the cohesion of which is further reinforced by interconnecting protein fibers. The shell has thin, partially organic inner and outer layers, but all but a small percentage of the whole is calcium carbonate, which is also the essence of chalk and limestone. The eggshell is porous, rather than solid, allowing the contents, the incipient chick, to "breathe."

Egg colors and patterns (see below) are also secreted from the walls of the shell gland. Two types of pigment—blue/green and red/brown/black—are produced, possibly from bile and blood, respectively. The blue/green coloration is deposited throughout the shell, and when present provides a uniform coloration (e.g., "robin's egg blue"), though varying in shade and intensity from species to species. The darker pigment is responsible for the patterns that ornament many eggs and can be deposited at various stages in the shell production. On the surface of the egg, these markings can range in color from pink to black and may also show through superficial layers of eggshell in muted tones. In some cases, the red/brown/black pigment appears uniformly in the surface layers (cuticle) of the egg, altering the underlying color: white becomes yellow or brown; blue becomes olive. In a few species a chalky coating is secreted as a final icing on the finished egg.

Superficial Characteristics of Eggs

SIZE. In general, the larger the bird, the larger its egg—though smaller birds generally produce eggs that are larger in proportion to the adult's body mass than is the case with larger birds. However, there are some anomalies (e.g., the Ruddy Duck is much smaller than the Canvasback but lays a larger egg). Kiwis lay enormous eggs relative to body size (25% of the female's total mass compared with the average range of 2%–11%). These large eggs represent a high percentage of female reproductive energy devoted to this stage of their offspring's life. The fossil eggs of the gigantic extinct Elephant Bird, *Aepyornis*, are as large as 14.5 × 9.5 inches (36.8 × 24.1 cm) and are thought to have weighed up to 27 pounds (12.27 kilograms). The largest egg of a living bird worldwide is that of the ostriches (the largest birds), both in diameter—average 7 × 5.5 inches (17.8 × 4 cm)—and in weight—3 pounds (1.4 kg). The egg of the Mute Swan at 4.5 × 2.9 inches (11.43 × 7.37 cm) is closely rivaled by the California Condor's—4.3 × 2.6 inches (10.9 × 6.6 cm).

Hummingbirds unquestionably lay the smallest eggs of any living birds, though which species holds the absolute world record is unresolved. The egg of the Bee Hummingbird (*Mellisoga helenae*—a Cuban

Kiwi egg

endemic and the world's smallest bird) is often cited: 0.45 × 0.32 inch (1.14 × 0.81 cm) or thereabouts, but lengths as short as 0.25 inch (0.64 cm) are reported for the family. In North America, the competition for smallest egg may be too close to call, given normal variation. The smallest hummingbird eggs weigh in at about 0.5 gram (0.0176 ounce).

Egg size may also vary within a given species according to: clutch size (more eggs = smaller ones); season (some songbirds lay slightly larger eggs in succeeding clutches during a given breeding season, while some seabirds' eggs are smaller on the average as the season advances); and age of the female (the eggs of some birds tend to be slightly larger as the female ages).

SHAPE. Most, but by no means all, eggs are "egg-shaped," that is, with one end slightly more pointed than the other. But the range of variation even within this readily visualized shape is broad, and several terminologies have been used to describe different shapes with varying degrees of exactness, for example, *elliptical, subelliptical, pyriform, ovate* (= oval = egg-shaped!), etc. Guillemots/murres and many shorebirds lay strongly pyriform (pear-shaped) eggs that have distinctly broad and pointed opposite ends. It is perhaps significant that pyriform objects roll in a circle, a useful adaptation for birds such as guillemots/murres, which lay their eggs on a bare, narrow cliff shelf. It is often suggested that among the shorebirds this shape is an adaptation to efficient incubation, since the four pyriform eggs that sandpipers and plovers usually lay form a neat, tight circle when clustered with their pointed ends together.

COLOR AND PATTERN. The normal ground colors of bird's eggs are white (i.e., without visible pigment), blue to green, or brown. Depending on where in the shell the pigment occurs, a wide range of hues and shades is possible. Though many eggs are "plain," the majority show some kind of marking, again in a form of brown pigment. The markings are classified by oölogists (people who study eggs) into categories such as scribbled, scrawled, speckled, spotted, and blotched, and such markings may occur in characteristic patterns. The eggs of many species, for example, are "wreathed," that is, have a concentrated circle of markings at the blunter end, which passes through the shell gland first and picks up the bulk of the pigment.

There is close color/pattern conformity in some bird families; for example, all North American owls have white or off-white, unmarked eggs. But there is also much variation, not only among species in a family but also among individuals of a species. In fact, *no two birds' eggs are exactly alike.*

Because the eggs of birds' nearest phylogenetic ancestors, the reptiles, are white, it is generally assumed that birds' eggs were originally white and evolved color and pattern in accordance with various survival pressures. Birds such as terns and plovers that nest in the open usually lay eggs that are cryptically colored and patterned to blend with the ground, making them less obvious to predators. The great individual variation in the patterns of guillemot/murres' eggs may have recognition value for individuals looking for their own eggs in a crowded colony. And some brightly colored eggs may advertise an unpleasant taste.

TEXTURE. Loon, swan and goose, alcid, chachalaca, stork, jaeger, many hawk, and some gull eggs have a *granular* texture (mostly rather fine). Duck eggs have a waxy or greasy coating, which may be a kind of water-proofing. The eggs of Andean and California Condors are finely pitted. Gannets, cormorants, darters, pelicans, flamingos, and anis lay eggs with a superficial white chalky coating, which tends to scratch or flake off at least partially during incubation. Except for the pelicans, the underlying shell color of these eggs is green/blue. Woodpeckers have notably smooth and shiny eggs.

Egg Collecting

Before it was prohibited by law, collecting birds' eggs was a popular hobby among men and boys in northern Europe and America. Good collections not only included a large species representation (with as many rarities as possible) but also displayed the greatest possible range of variation in the eggs of a given species and often different-sized clutches of the same species. Clandestine egg collectors are still a bane to conservation authorities in parts of Europe (especially Britain), but on the whole the passion for the activity seems largely to have died out in North America. An exception was the theft in 1981 of a nest and eggs of Ross's Gull from its then newly established breeding station at Churchill, Manitoba. The culprit, reputed to be Austrian or German, apparently cut the entire nest and the grassy tussock on which it was placed from the tundra with a sharp tool between 7 p.m. and 3 a.m., managing to elude a guard posted to protect the site. It is estimated that the eggs would bring as much as $10,000 to $20,000 in the oölogical underground. The whereabouts of the nest and eggs remain unknown.

Since the rarest birds inevitably lay the rarest eggs, we can be grateful that the thrill of risking jail and a stiff fine for stealing a bird's egg that can be admired by only a handful of other egg-loving criminals appeals only to a kinky few.

Eleanora of Arborea (1347–1404)

Arborea was one of four independent "judicates" (essentially kingdoms) into which the island of Sardinia was divided in the Middle Ages, between the ninth and fifteenth centuries. Each was ruled by a judge with the authority of a king. Marianus IV, the only Sardinian ruler to be known as "the Great," reigned over an era of "splendor" in Arborea from 1353 to 1375. He was succeeded by his son Ugone (Hugh) III, who was killed in a political conspiracy in 1383 along with his daughter, leaving no issue. There was a move to declare a republic, but Eleanora, Hugh's eldest sister, claimed the judgeship on behalf of her sons. She became one of the last, most powerful, and most effective of Arborea's leaders, widely regarded as Sardinia's greatest heroine due to a code of laws she enacted that broadened the rights of citizens in matters of sex and inheritance, among other issues.

Eleanora and other family members were ardent falconers, and in 1392 under her code of laws, she provided for the protection of birds' nests from illegal hunting for the first time in history. Her ornithological distinction is preserved in Eleanora's Falcon (*Falco eleonorae*), which nests colonially, mainly on Mediterranean coastal cliffs and islands—including Sardinia.

Evolution of Birdlife

Even a cursory inspection of our planet's present-day birdlife reveals an astonishing diversity of avian forms: albatrosses and penguins, condors and ibises, kingfishers and nightjars, dippers and hummingbirds—roughly 10,721 species, each highly adapted to a specialized role in the earthly ecosystem. It is even more of a wonder then that this represents at most a tenth of the bird species that have soared, waded, and flitted through habitats that no human eyes ever saw and that entire ancient avifaunas arose and disappeared before our species emerged from apedom a mere 6 million years ago. We owe our ever-increasing knowledge of these previous worlds (1) to the lucky natural accident that extinct animal forms are preserved in stone and asphalt; (2) to the ingenuity of the scientists who have figured out how to date rocks accurately and even tell time using molecular clocks; and (3) to Mr. Darwin and others who showed us how these fossils could be related to each other.

Doubtless many more insights about prehistoric birdlife lie buried than we have yet unearthed. Yet the fragments that we have managed to discover and fit into their places in the evolutionary puzzle have proven sufficiently fascinating, including some remarkable discoveries within the last decade. This entry can do no more than sketch the contours of early birdlife and its origins. Fortunately, excellent comprehensive texts on the subject are now readily available.

Reading the Rocks

PERSPECTIVE. Since our only physical evidence of prehistoric birdlife is fossilized remains, we should remember that this source of data does not present anything like a thorough and balanced overview of the past. For one thing, there isn't very much of it. Then, as now, the fragile bird skeletons that evolved as concomitants to flight were pulverized in predators' digestive tracts or scattered and broken by scavengers and moving water far more often than they fell intact into preservative sediment or tar pits. Furthermore, fossils tend to present a habitat bias. Lake-bed sediments, after all, can be expected to contain a certain number of fossilized lake birds, but not to yield many clues as to what kinds of birds inhabited the lakeside forests. Finally, it should be noted that paleornithologists are themselves "rare birds," with the result that old bird bones have been looked for and studied less than either modern birds or extinct lizards. However exciting new fossils are to our imaginations, we must keep in mind that our present picture of prehistory is like an obliterated fresco in which we have managed to replace a few randomly unearthed fragments.

THE ANCESTRY OF BIRDS. That birds evolved from reptiles has not been disputed by science since the time of Darwin; birds and reptiles share many traits, and their relationship is literally embodied in the famous "lizard bird," *Archaeopteryx* (see below).

But a continuing argument about the descent of birds concerns their immediate ancestors. School children now know that birds are the living descendants of dinosaurs. But which dinosaurs? The current consensus among scientists holds that based on an analysis of similar traits, birds evolved directly from coelurosaurs—a type of theropod dinosaur. Theropods include not only the fearsome *Tyrannosaurus* but, more pertinently, the small, agile, fast-running predators, such as *Velociraptor*. The fact that theropods and *Archaeopteryx* fossils share the same geological strata seems to support this view. However, the paleornithologist Alan Feduccia

Fossil of *Archaeopteryx*

and others argue credibly that the apparent similarities between early birds and the theropods are either nonexistent when examined closely (e.g., the perceived similarity in bird and theropod three-digit "hands") or result from CONVERGENT EVOLUTION. This camp holds that birds sprang directly from *thecodont* reptiles lower on the crocodile branch of the vertebrate tree and cites many "pre-avian" traits in support of this theory. Why, then, the theropodists ask, is there no lizard-bird in the fossil record for another 90 million years? It is a classic scientific debate replete with both brilliant argument and personal insults, which will probably be resolved only with the discovery of new fossils.

EVOLUTION OF FEATHERS AND FLIGHT. The close relationships between reptilian scales and feathers and between modern bird feathers and flight are undisputed. The scales of some of the fossil thecodonts seem to bear the pattern of a feather impressed on their surface, and the processes of scale and feather formation are very similar in modern reptiles and birds. Likewise, the adaptive elegance with which feathers suit flight in present-day birds seems obvious, though it is equally clear that feathers also serve other functions. However, reconstructing the step-by-step transition from scale to feather and earthbound to airborne remains an even more speculative—and controversial—undertaking than exploring the branches of the avian family tree.

If you think of the typical reptilian scale, a basic body-armoring device, it isn't difficult to imagine potential evolutionary improvements: If the scale is split at angles along the midrib, it becomes more flexible; if it is frayed, it could provide more insulation and therefore the ability to be useful in extremes of heat and cold; if the structure becomes thinner and more open, weight is reduced and agility increased; if limb, tail, or back scales are elongated, they can improve balance and maneuverability and permit parachuting or gliding; and all of the preceding modifications could also function as mechanisms for display, making sounds, or other secondary uses. All of these are, of course, speculative mutations that would have conferred selective advantages on scaly animals on their way to being feathered ones. But was flight the irresistible advantage that turned scales into feathers or, as herpetologists tend to think, were feathers initially a more sophisticated means of controlling body temperature that later came in handy as reptiles began taking to the air? Current thinking seems to favor the latter, as recent analysis of Chinese fossils argues that dinosaurs evolved feathers long before what we would call birds appeared.

The debate about the origins of powered avian flight breaks out pretty much along thecodont/theropod lines into "trees-down" and "ground-up" camps. The thecodonts took many forms, and we know from fossils that some were small, light, arboreal species with greatly elongated scales; some also gave rise eventually to flying reptiles such as *Pteranodon*. These might have begun their evolution toward flight with just enough extra scale extension to slow a nervy leap from one tree to another, called "parachuting." The next stage for an "avimorph thecodont" would have been the development of a broader airfoil, giving the ability to glide some distance after launching from a perch and to control the quality of the ride. The 2-inch-long *Longisquama* ("long-scale") is the only reptile known so far to have had scales that may be intermediate with feathers; it had a remarkable double fan of scales extending from its back that it could apparently unfold like a butterfly's wings and glide—though not flap—around the Triassic forests.

The so-called cursorial or ground-up theory evokes the image of the nimble theropod racing around the desert on its hind legs catching whatever it can grab with its forelegs. Elongation and fraying of these fore-limbs would have nothing to do with flight initially, but would simply allow higher leaps after prey with better control and/or the ability to catch more prey with greater ease using the winglike arms. As the wings and tail broadened, leaps could become glides, followed eventually by powered (flapping) flight. One thecodon-tian objection to this scenario is that the fossil thero-pods that we know of so far do not show any tendency toward the elongation of the forelimbs and shifting of weight forward that are required by aerodynamic prin-ciples to get a running dinosaur into the air.

So perfectly does *Archaeopteryx* embody the emer-gence of a reptile into bird-dom that it is claimed as evidence by both sides in the origins-of-feathered-flight controversy. Tree-downers say its ancestors were arbo-real reptiles that evolved feathered wings and tails as gliding devices; they cite its pronounced backward-pointing toe (hallux) as proof of its tree-perching pro-clivity. Ground-uppers argue that it supplies the nec-essary link between the running, leaping, bug-catching theropods and the birds they became and that it climbed into the trees from its original terrestrial habitat.

The Fossil Record

The following overview attempts to sketch the broad shape of the fossil record to date and includes brief de-scriptions of those extinct bird species whose skeletons have excited the most wonder.

ARCHAEOPTERYX AND ITS RIVALS. Though many significant discoveries have been made in recent years, it is probably still fair to say that the most important bird fossil that has yet been found is that of *Archaeopteryx lithographica*, seven specimens of which, including complete skeletons, have been liberated from late (Upper) Jurassic limestone sediments in Bavaria, the first of them in 1861. In addition to being the first-known feathered animal, the significance of *Archaeopteryx* lies in the fact that its skeleton is more that of a reptile than a modern bird. It has, in fact, been described by more than one authority as a small (crow-sized) dinosaur. And while it now resides officially in an infraclass (or subclass) of its own, Archaeornithes ("ancient birds"), it is described by many authorities as a feathered reptile. Coming as it did just two years after the publication of Darwin's *On the Origin of Species by Means of Natural Selection* (1859), this "missing link" was celebrated not just as the earliest known bird, but for many as conclusive proof of the validity of Darwin's theory. For here was an organism clearly in transition between two well-defined groups of animals, reptiles and birds.

The *Archaeopteryx* fossils are dated at 135–155 million years before the present, in the middle of the Mesozoic era, the so-called Age of Reptiles. Its habitat appears to have been a dry tropical shore with highly saline lagoons not unlike the present-day coast of northeastern Venezuela. The vegetation was probably dominated by cycads, including palmlike tree species, and the air was apparently filled with several species of flying reptiles (pterosaurs). Regardless of whether *Archaeopteryx* was descended from arboreal or terrestrial ancestors, authorities generally agree that it lived in trees and possibly behaved rather like modern guans, walking along branches, and gliding from tree to tree; the structure of its skeleton and feathers suggest that it may have been capable of some degree of powered flight. It also bore long claws at the bend of the wing, which it presumably used for crawling around the tropical canopies.

The fame and detail of the *Archaeopteryx* fossils make them, as it were, the mark to beat among paleontologists in search of a new "ur-bird," and several candidates have been promoted. Arguably the most credible rival to *Archaeopteryx* is the wonderfully named *Confusiusornis sanctus* described in 1995 among many other bird fossils from lake-bed sediments in northern and eastern China by Hou Lian-Hai and Zhou Zhonge of the Academia Sinica in Beijing. Plausibly dated as contemporaneous with *Archaeopteryx* in the late Jurassic, *Confusiusornis* combines many of the former's primitive traits with more modern features. It is toothless and one of its fossils shows evidence of a feathered body.

CRETACEOUS BIRDS. Leaving the Jurassic, we enter a period of geological time during which life on Earth and indeed the surface of the Earth itself underwent a sequence of remarkable changes. During the 80 million years of the Cretaceous period from 146 to 65 million years before the present, existing continental land masses moved apart significantly—a major influence on species distribution; flowering plants appeared and broad-leaved forests gained dominance as early conifers declined; the great dinosaurs flourished worldwide and then declined, and marsupial and early placental mammals evolved. At the end of the Cretaceous a meteorite (or perhaps several meteorites) smashing into the Earth at 50,000 miles per hour and/or a series of volcanic eruptions created a global cataclysm involving fires and tidal waves and an encompassing cloud of dust that either cooled the planet by blocking the sun's rays or warmed it by creating a greenhouse effect. The result was a period of mass extinctions during which the dinosaurs and many other species that prospered during the Cretaceous disappeared altogether from the Earth's biota. The event or series of events that changed the world is detectable in the geological strata as a thin line of dust that settled out of the atmosphere as it cleared. Known as the K-T boundary, it marks the border between the Cretaceous period and the Tertiary period that followed and also an invisible wall in time between two biologically different worlds.

Early Modern Birds
Perhaps the best-known examples of Cretaceous birdlife were two groups of toothed seabirds represented most prominently by loonlike species in the genus *Hesperornis* ("western bird") and the ternlike members of the genus *Icthyornis* ("fish bird"). Both types appear to have ranged worldwide or nearly so and to have evolved into a variety of species. The hesperornids, of which 13 species are known, were large (up to 5 feet long), heavy-bodied diving birds with only vestigial wings and large swimming feet located at the extreme stern end as in present-day loons and grebes; they had lobed toes not unlike those of modern coots and phalaropes; closely

related species in the genus *Baptornis* ("dunking bird") were similar. We know from coprolites (fossil feces) found with some *Baptornis* remains that these birds were fish eaters.

Judging from skeletal features, *Ichthyornis*—and the similar *Apatornis*—were strong-flying predaceous species that foraged over the great inland seas of North America and beyond during the late Cretaceous. *Ambiortus dementjevi* recovered from Mongolian lake-bed sediments is very similar anatomically to *Icthyornis* and *Apatornis*, but it was a land bird and, dated in the early Cretaceous, is presently the oldest known ornithurine (modern) bird.

There are four things to remember about these birds: (1) They are structurally much closer to modern birds than to *Archaeopteryx*; (2) despite this they do retain one distinctly reptilian feature: rooted teeth set in sockets; (3) though "modern" and "loonlike" (or "ternlike"), they are not phylogenetically related to the loons and terns that evolved in the Tertiary; and (4) they were diverse, very common, and widespread during much of the Cretaceous, declined gradually toward the end of the period (because the inland seas were shrinking?), and then disappeared completely along with the dinosaurs in the K-T extinction.

The Transitional Shorebirds

Since no toothed birds, or early modern (ornithurine) birds are known to have survived the Cretaceous extinctions, and since no representative of a modern family is known for certain from the Cretaceous, we have a problem: Where did the hugely diverse Tertiary avifauna come from? What was the genetic base on which it was built? One way to answer this is simply to say we don't know yet and encourage the paleornithologists to keep digging. However, using the DNA of recent birds as a "molecular clock," some ornithologists have concluded that more than 20 orders that evolved much earlier in the Cretaceous than any fossil evidence confirms made it across the K-T boundary. While these findings have been disputed on a number of grounds, it is certainly possible that the discovery of more fossils will change our mind about extinction and survival at the end of the Cretaceous. For now, the main working hypothesis for this conundrum involves a few known fossil species that may have slipped through the K-T barrier. Among these "transitional shorebirds" are species once identified with modern loons, tubenoses, cormorants, flamingos, and rails, but now thought to belong in the Charadriiformes, the order containing modern sandpipers, gulls, and related forms. Unfortunately, these fossils are either extremely fragmentary or uncertainly dated or both, and no authority is comfortable making such an important call until there is better evidence. An intriguing argument put forth in favor of the shorebird hypothesis is that these birds might have been comparatively well adapted to survive the effects of the K-T catastrophe. If the Earth passed through a period when photosynthesis was impaired because sunlight was occluded, many plant-dependent species and their predators would have rapidly perished, while species like sandpipers and sandgrouse that can make do with seeds and invertebrate eggs would have been better able to squeak through.

THE TERTIARY "EXPLOSION." If the Cretaceous can be said to have ended with a bang, the early Tertiary had a certain explosive quality as well, albeit of a different kind. We tend to think of evolution as an almost painfully gradual process with small incremental changes taking place over millions of years. But the fossil record shows that at least in some instances an evolutionary "opportunity" allows certain groups of organisms to diversify rapidly (by geological standards) to fill newly available ecological niches. The dawn that eventually came following the long night of the K-T transition revealed a world with many fewer species, waiting, as it were, to be reinhabited by new assemblages of animal life. The gap left by the dinosaurs made room for the rise of some intimidatingly large birds (see GIANT BIRDS) and, eventually, a new fauna of large mammals. And whatever smaller "transitional" birds made it into the Tertiary had the place to themselves and underwent what is invariably referred to in the literature as an explosive radiation that "invented" the modern avifauna in only a few million years. By the Eocene epoch, just 10 million years into the new period, all modern orders of birds except the perching birds (passerines) had evolved—and left sample skeletons to prove it. The passerines, which were still nowhere to be seen in the Oligocene 30–40 million years into the Tertiary, experienced their own dynamic radiation in the Miocene, resulting in the advent of more than 50% of the modern avifauna. Almost as fascinating as this Tertiary burst of evolutionary creativity is the fact that once the modern genera of birds had emerged, the expansion all but stopped, nature confining herself to fine-tuning at the species level up to the present day. Time-traveling birdwatchers finding themselves 10 million years back in the Miocene would recognize the great majority of genera they encountered, though a lot of the species might be puzzling. (Birders should note that when this option becomes available, the destination to book is the Pliocene—just 5 million years or so back when the Earth's birdlife probably reached its greatest diversity, perhaps with as many as 150,000 potential "ticks.")

OF ICE AND MEN. In the Quaternary period (from the beginning of the Pleistocene epoch to the present), the gradual cooling of the Earth's climate begun in the Tertiary culminated in a series of glacial advances and retreats across the Northern Hemisphere. The advent of a bipolar, strongly seasonal climate subject to periodic "ice ages" had a pervasive influence on the distribution of the Earth's biota, including its birdlife. Among other effects it created zones of extreme heat and cold, and of dryness and wetness that in many cases acted as ecological isolating mechanisms promoting speciation.

As glaciers advanced, they pushed tropical habitats and their resident species toward the equator, marooned species on mountaintops, and were probably a major factor in the evolution of migratory patterns. While it is reasonable to suspect that these inexorable ice sheets hastened the doom of the many bird species that vanished during the Pleistocene, one of the glaciers' most significant impacts on Pleistocene life-forms may have been an indirect one. During the periods of maximum glaciation, much of the planet's water turned to ice, resulting in a drop in sea level amounting to hundreds of feet. This created land bridges spanning continents that had formerly been isolated from each other. About 14,000 years ago human beings walked from what is now Siberia across the Bering land bridge and into the New World for the first time. It took us only 2,000 years more to reach the tip of South America.

We arrived on a continent populated by the so-called Pleistocene megafauna, vast numbers of elephants, horses, and other herbivores, many predators, such as the saber-toothed cats (*Smilodon* and other genera) and the Dire Wolf (*Canus dirus*), and a large community of scavengers, including the vulturine teratorns (see GIANT BIRDS), and another—relatively puny—scavenger that prospered during the Pleistocene, the California Condor. Some students of the Pleistocene megafauna and its demise have recently argued persuasively that the collapse of this ecological community not long after the arrival of people in its midst may not be simply coincidental. Alas, this would be consistent with the fate of large animals and especially large birds—elephant birds, moas, dodos, Great Auks—that quickly follows their first meeting with our species.

No new species of birds are thought to have evolved since the end of the Pleistocene and since about 80 have become extinct within the last 300 years, the decline of bird diversity begun around the time of the last glacial interval continues—largely under human auspices.

Fallout

Birders' jargon for the sudden landing of large numbers of migratory birds, typically as the result of weather conditions such as fog or the arrival of a front coming from the opposite direction of the migration route. The overnight appearance of a "wave" of migrants returning over water from tropical wintering grounds and concentrating on the first available coastal habitat is one form of fallout. But the appearance of smaller numbers of rarities becoming disoriented in bad weather, such as tropical storms, or overshooting their intended destinations and making an emergency landfall outside their normal range also qualifies.

Fiction

Excepting the odd parrot on a sea captain's shoulder or caged canary used as a prop, birds in fiction usually serve as background scenery and mood-setting devices. Novelists with an eye for nature or a fondness for outdoor action frequently conjure the seacoast with some gulls wheeling and mewing; it is very difficult to describe spring convincingly without a certain amount of exuberant birdsong; and some ecstatic avian caroling from field or forest is as indispensable as the perfume of wildflowers to love alfresco. There is also, evidently, a strong compunction, when writing about tropical landscapes, to include some "strange" and/or colorful birds for the sake of verisimilitude.

Such uses of birds, if well done, may enhance a reading experience for someone who knows something about birdlife, or it may inspire a chuckle if the author has skipped her ornithological homework. Overall, however, the impact of birds in fiction must be accounted as negligible.

A few oddities where birds assume a more prominent role than usual may be mentioned:

Gulliver's Travels (1726) by Jonathan Swift. The hero is "boxed" by a Brobdingnagian Linnet "somewhat larger than an English Swan" and in the same country of giants, he narrowly escapes a "kite," which sounds suspiciously like a kestrel.

The Narrative of Arthur Gordon Pym of Nantucket (1838) by Edgar Allen Poe includes some fairly detailed and (on the whole) accurate accounts of South Polar oceanic birds, which add a suitably bizarre touch to the later chapters. Poe also invents a few species of his own, for example, a "Black Gannet."

Green Mansions (1904) by W. H. Hudson. Some touches involving tropical forest birds help the reader pass through the soggy romantic wilderness. Hudson's reminiscences of his naturalist boyhood on the Argentine pampas (especially his 1918 *Far Away and Long Ago*) are far superior, from both an ornithological and a literary perspective.

An American Tragedy (1925) by Theodore Dreiser. The Adirondack wilderness setting of the famous drowning scene (Book II, Chapter 47) is alive with birds that either call ominously or provide a counterpoint to the grim occasion with their fine songs or bright plumage. Though Dreiser uses his avian images to good effect, he is an indifferent ornithologist. But his inclusion of a Caribbean endemic (Yellow-shouldered Blackbird) in the avifauna of upstate New York is more than compensated by that sinister corvid, the "weir-weir," which hops about the branches of dead trees during the murder, going "Kit, kit, kit Ca-a-a-ah!"

Birds of America (1971) by Mary McCarthy. The central character is a young birdwatcher (= sensitive), and

Mail delivery by
Tawny Owl
at Hogwarts

the novel is about a phase of his late adolescence. It gets some good mileage out of the "personalities" and habits of a number of bird species.

Peter Matthiessen, a keen birdwatcher and brilliant chronicler of the natural world, often includes avian images in his fiction, including the epic *Shadow Country* (2008), set in the Florida Everglades.

J.R.R. Tolkien makes excellent use of magisterial eagles and venerable, watchful ravens, not to mention the wise old thrush in *The Hobbit*.

And, of course, the owls of Hogwarts, including Harry Potter's messenger Snowy Owl, Ron Weasley's pet Scops Owl, Draco Malfoy's "familiar," an Eagle Owl, and swarms of Tawny Owls delivering the mail.

The above does not, of course, pretend to be a comprehensive survey of birds in fiction. Perhaps this may inspire someone to make one.

Flightlessness

It is tempting to liken a bird that can't fly to a fish that can't swim, but this turns out to be a poor analogy. With very few exceptions, such as the swifts, birds tend to be at least as well adapted to life on the ground or in the water as they are to an aerial existence. Some birds, gulls, for example, seem equally adept at all three. It is not very surprising therefore to find that there have always been and continue to be birds that have evolved terrestrial lifestyles at the expense of their power of flight. And it is now generally believed that all flightless birds have descended from flying ancestors.

Though doubtless a precursor of stronger-flying forms, that early ancestor to birds, *Archaeopteryx*, was a poor flier whose wings mastered the air only to the extent that they enabled their owner to glide from tree to tree, much as a flying squirrel does today. Later diving birds in the genus *Hesperornis* of the Cretaceous period had only the tiniest of vestigial wings and passed their lives on and under the water, while strong-flying reptiles flapped and soared overhead. Terrestrial flightless birds often show a conspicuous protective adaptation that compensates for their inability to escape into the air. In many cases this adaptation takes the form of great size (see GIANT BIRDS), though the kiwis of New Zealand rely on secretive nocturnal behavior rather than on size for protection.

On isolated islands or lakes, species of flightless grebes, cormorants, ducks, and rails evolved in the absence of predators and consequently lacked any effective defense when man and his attendant rats, pigs, cats, and dogs disembarked in their domains. Many unique species, for example, the Dodo and Great Auk, have ceased to exist through a combination of human callousness and their inability to fly from it, and, almost without exception, those that survive are severely threatened. Given its obvious disadvantages, one is compelled to ask what the compensations of flightlessness might be. In addition to the circumstance that the ability to fly simply became unnecessary for nonmigratory species on predator-free islands, it has also been suggested, beginning with Darwin, that flightlessness (in insects as well as in birds) might decrease the risk of getting blown out to sea.

Unique among modern flightless birds are the penguins, which, unlike the ostriches and other ratites, did not lose the power of flight but rather adapted it to swimming. The term "ratite" implies the lack of a keel on the breastbone, to which the flight (pectoral) muscles of flying birds attach. Penguins have both keel and flight muscles, but their wings have become flippers, and they "fly" through the water rather than in the air.

If we survey the present world avifauna, it becomes clear that the ability to fly is not represented in a simple "have" or "have not" fashion but is highly variable among different species. Some birds, for example, albatrosses, vultures, and alcids, are powerful fliers under the proper conditions but are unable to get off the ground in others. Many species fly well enough but do so relatively infrequently, spending most of their energy in the water (grebes) or on the ground (many gallinaceous birds). Even within a single family, great variation in flying ability may be apparent. Many cuckoo species are graceful fliers and long-distance migrants, yet the related anis seem barely able to cross a road despite a furious flapping of their short wings followed by a long *Archaeopteryx*-like glide; and those giant cuckoos the roadrunners, though apparently more capable than the anis, almost never take to the air, but prefer to run—even for their lives.

The only passerine bird species suspected of being flightless was the Stephen Island Wren, a member of a small family of birds (Acanthisittidae) endemic to New Zealand. It reputedly ran around on the ground like a mouse, but its habits were little known and it became extinct soon after it was discovered in 1894; its wing structure suggests that it may have been capable of weak flight.

Gause's Rule (a.k.a. the principle of competitive exclusion)

Holds that no two (or more) organisms with exactly the same ecological requirements can coexist in the same environment permanently, since the inevitable competition will always favor one species over the other(s). Therefore, one normally finds birds (and other animals) occupying different ecological "niches" in any given habitat.

Giant Birds

Among the most impressive birds that ever lived were the species of *Gastornis* (formerly *Diatryma*), the largest of which reached nearly 7 feet in height and weighed close to 400 pounds. The remains of these and similar flightless giants have been found in Paleocene and early Eocene deposits in New Mexico, Colorado, Wyoming, New Jersey, and the Canadian high Arctic as well as Europe and may be relatives of cranes or geese. Because of their massive bills and powerful build, the prevailing view has been that diatrymas were ferocious predators that ran down their prey in open country. However, an alternate theory suggests that the great bill

Gastornis gigantea with "prey"

was a cropping device and that the huge birds wandered flood plain shrublands eating bushes. The similar but more agile phorusrhacids of South America are still believed to have been swift-running predators. These feathered monsters went extinct along with much of the rest of the South American fauna when the Panamanian land bridge permitted the southward invasion of North American mammalian predators. Before this happened, however, a 10-foot-tall phorusrhacid, *Titanis walleri*, found its way north to Florida, where it probably fed on prey up to the size of small deer.

Another gigantic avian relic from this period is a seabird from California with a wingspan as great as 16 feet (modern-day Wandering and Royal Albatrosses have a maximum documented wingspan of between 11 and 12 feet, the greatest of any living bird species); of more interest to ornithologists are the toothlike structures that arise out of the jawbones, giving the species its Latin name, *Osteodontornis* ("bone-tooth bird"). Finally, there were the great vulturine teratorns, including the North American *Teratornis incredibilis*, which had a 17–19-foot wingspan and weighed more than 40 pounds; its cousin from Argentina, *Argentavis magnificens*, weighing nearly 160 pounds with a wingspan of up to 25 feet, was the largest flying bird we know of.

The much more recent Elephant Bird of Madagascar is thought to have weighed half a ton, and the largest of the giant moas of Australasia stood as tall as 13 feet—proportions that would inhibit most predators, though there was a moa-eating eagle in New Zealand.

Of course, most of the living ratite birds, the ostriches, Emu, cassowaries, and rheas, are also very large (see RATITES).

Goatsucker

A widely used English name, especially in North America, for most members of the nightjar family Caprimulgidae (literally "goat-milkers"). The superstition that members of this family suck at the teats of goats and other livestock at night, ultimately causing them to waste away, goes back at least to Aristotle's time and remains prevalent in many cultures. The nightjars' peculiar appearance, nocturnal habits, and eerie calls doubtless reinforce the myth.

Griscom, Ludlow (1890–1959)

The patron saint of modern American birdwatching. As an ornithologist at the American Museum of Natural History in New York and at Harvard, Griscom's main professional contribution was the elucidation of the Mexican and Central American avifaunas. But the achievement of his life was to show that "birding" did not need to be practiced with a gun—except in cases where field identification is impossible and a specimen is required for the record—and that the great majority of birds can be identified with absolute accuracy by a practiced and intelligent observer. Griscom's brilliance in the field, his eloquent enthusiasm for the "sport" of field ornithology, and his large circle of protégés (including Roger Tory PETERSON) were responsible to a large degree for spawning the legions of present-day birdwatchers and their need for good optical equipment and comprehensive field guides. Griscom's love of fieldwork is reflected in his best-known works, which were mainly distributional in nature.

Guano

Originally the excrement of the Guanay Cormorant (*Phalacrocorax bougainvillii*) and other island-nesting seabirds off the west coast of South America. The guano, which accumulates to significant depths at the largest colonies, has been commercially "mined" for fertilizer and gunpowder and was once Peru's chief

Guanay Cormorant

export and source of income. By the end of the nineteenth century, however, reckless exploitation had depleted supplies of guano built up over the centuries. More recently, an attempt has been made to extract the guano in equilibrium with its accumulation.

The excrement of other species of colonial birds has been exploited elsewhere in the world, but the food preferences of the seabirds and the arid climate of the west coast of South America are ideal for the production of nitrogen-rich guano of commercial scope.

The term is now used more broadly for other forms of excrement, for example, "bat guano."

See also POOP, ETC.

Grey-headed Kingfisher
(*Halcyon leucocephala*)
of Sub-Saharan Africa

Halcyon

A literary name for kingfisher, and, by a combination of fanciful metaphor and bad ornithology, an adjective meaning calm, peaceful, for example, "halcyon days."

A minor Greek goddess, Halcyone threw herself into the sea out of grief for her drowned husband, Ceyx. (Both *Halcyon* and *Ceyx* have been adopted as names of kingfisher genera.) Taking pity, the gods turned them both into kingfishers. From this legend arose the belief that kingfishers nest on the surface of the sea for a period of two weeks near the winter solstice and have the power to calm the waves to facilitate the incubation process.

For the record, the majority of the world's 95 species of kingfishers do not fish, and most of those that do avoid saltwater; all nest in earthen burrows.

Left-footed parakeet

Handedness/Footedness

Perhaps it is not surprising that few studies have been undertaken to determine to what degree birds are left-handed or right-handed (or to be precise in this case left- or right-*footed*). There is some evidence that a majority of individuals of at least some parrot, hawk, and owl species are lefties. On the other hand, the feral pigeon (and presumably the wild Rock Pigeons from which they are descended?) may be largely right-footed. It has also been shown that in some instances the preferred foot is slightly longer than the other and that some individual birds appear to be ambidextrous.

Hawkwatching

This form of birdwatching has taken on a life of its own in recent decades, owing its popularity, no doubt, to the inherent glamour of birds of prey and to the impressive concentrations in which they can be seen during migration. These aerial rivers and eddies of raptors occur because migratory soaring birds depend on air currents such as thermals and updrafts to facilitate their journeys (see SOARING) and seek out these conditions where they occur in nature. These birds also migrate by day in large numbers and during a narrow window of time. The birdwatcher who visits the right mountaintop, ridge, lake shore, or geographical cul-de-sac (think Cape May, New Jersey) on the right day in spring or fall will witness one of nature's most thrilling spectacles. The ur-hawkwatching locale in North America is Hawk Mountain, Pennsylvania, and a few other traditional spots have attracted birders in season for decades. Fairly recently, however, a new breed of hawkwatcher has actively sought out undiscovered watch points— finding hundreds of "new" ones worldwide—and has also started counting the raptors in a methodical way to foster their conservation.

There is now a small library of books on the subject and a number of organizations worldwide, such as the Hawk Migration Association of North America (HMANA), with its mission "to conserve raptor populations through scientific study, enjoyment, and appreciation of raptor migration."

Hearing

Birds are among the most vocal of animals, and it is therefore not surprising that they possess good hearing abilities, at least within their species' vocal range. Courtship and territorial songs, vocal signals between parents and young, alarm notes, threat sounds, calls among flocking birds, and sounds made by predators are important elements in the lives of birds and would be useless, of course, unless they could be heard.

THE AVIAN EAR. Birds lack the fleshy external sound-catching appendages so prominent in mammals, and their ear holes are usually totally concealed under the feathers called ear coverts (auriculars); these feathers apparently serve to protect the inner ear from wind disturbance in flight while permitting the passage of sound. In penguins and other diving birds the feathers covering the ears are thickened and act as earplugs. The skin and muscle that surrounds the outer ear is adapted for specialized functions in different groups of birds. Deep divers can close a flap extending from the rear rim of the ear. And owls have flaps of skin (opercula) in front of and behind the ear holes that can alter the size and orientation of the opening to focus and enhance sound perception. Owl ears are also unique (as far as known) in being asymmetrical both in position (one higher than the other) and in internal structure, adaptations for finding prey in the dark.

WHAT BIRDS CAN HEAR. To understand the useful comparison between human hearing and that of birds, the reader will recall that sound is recorded as the rate at which sound vibrations pass through the air, called cycles per second (cps or Hertz, abbreviated Hz). The more vibrations or cycles per second, the higher the frequency or pitch. The normal human ear hears sounds between about 20 and 17,000 Hz (23,000 maximum), and dogs can hear sounds up to 45,000 Hz. The known hearing range for all birds is between 34 and 29,000 Hz. *However*, the range of any single bird species is significantly less than ours: we hear about nine octaves; birds average about five. Nor is the hearing ability of birds especially acute; humans can generally pick up fainter sounds than most birds across the frequency spectrum.

As logic suggests, bird species tend to hear about the same range of sounds they can produce. Many small songbird species can sing and hear sounds of higher frequency than we are capable of hearing. However, these same birds miss several lower octaves that we hear easily. For example, these avian tenors and sopranos often cannot hear the relatively low-frequency human voice, so that loquacious birdwatchers usually disturb their fellow birders more than their quarry.

At least some birds can hear "faster" than we can. Slowing down a recorded bird song often reveals notes we couldn't hear at normal speed, though birds that mimic other birds (see SONG) have been shown to include these fast notes in their imitations.

USES. Birds use their hearing not only in communicating but also in locating food. In 1962 biologist Roger Payne revealed how Barn Owls can zero in on mice in the dark by precisely locating the position of their squeaks and scuffles. There is some evidence that robins, plovers, and other birds that hunt over the ground for invertebrates can hear their prey as it moves under the surface—and that woodpeckers can hear grubs and other wood-inhabiting insects moving in bark and trunks.

It has been shown that domestic pigeons can detect "infrasounds" (frequencies below 20 Hz), which include vibrations made by tectonic disturbances and more mundane sounds like the rumble of waves breaking along a coast—possibly at great distances. This would explain instances of birds "predicting" earthquakes, for example, roosters that crow in alarm before people are able to feel earth tremors. Birds cannot, however, hear the ultrasonic (very high frequency) sounds on which bats depend.

Echolocation, highly developed in bats, has been verified in a few species of cave-dwelling birds in South America (Oilbird) and Asia (swiftlets). However, the clicks that these birds bounce off their home caves to keep from bumping into the walls are well within normal hearing frequencies and thus much cruder than the ultrasonic signals used by bats.

Hemenway, Harriet Lawrence (1858–1960)

Founder, along with her cousin and neighbor, Minna Hall, of the Massachusetts Audubon Society and by extension the wildlife conservation movement in the United States. By birth and marriage, Mrs. Hemenway was a member of the socially prominent and wealthy class of Boston citizens known as Brahmins. On a January morning in 1896, she read in the newspaper of the slaughter of millions of egrets for the purpose of using their nuptial plumes to decorate fashionable women's hats. Since the most valuable plumes appear only in the nesting season, vast numbers of young birds were left to die when the parent birds were killed and plucked. At the time, the use of the feathers—and sometimes entire birds—of more than 50 species was a mainstay of the immensely profitable millinery trade.

The two women resolved to end the practice, which had already decimated egret colonies in Florida and tern colonies in New England. Consulting *The Boston Blue Book*, a listing of well-to-do families, they held a number of tea parties, inviting their fellow patricians—most of whom were likely plume wearers—to forswear the practice and to "work to discourage the buying or wearing of feathers and to otherwise further the

protection of native birds." Within the year, they had recruited more than 900 women to their cause; gained the support of Boston's highest-ranking scientists and bird conservation advocates; and established the Massachusetts Audubon Society, still the largest independent state Audubon Society. A key initial goal was to inspire the creation of similar organizations in other states, and by 1897, Audubon societies had been founded in Pennsylvania, New York, Maine, Colorado, and the District of Columbia; these soon coalesced into a national "association," which eventually became the National Audubon Society. Also in 1897, Massachusetts passed a law prohibiting the trade in wild bird feathers.

It will surprise no one to learn that the powerful feather barons who became immensely wealthy from the plume trade and their supporters in various legislatures were not complacent at the threat of losing their ill-gotten profits. At first, they mocked the incipient movement to protect birdlife as a "ladies club" and wondered why anyone should be concerned about "long-legged, long-beaked, long-necked bird(s) that live in swamps and eat tadpoles?" as U.S. Senator James A. Reid of Missouri asked from the Senate floor in 1913. "Why should we worry ourselves into a frenzy," he continued, "because some lady adorns her hat with one of its feathers, which appears to be the only use it has?"

But the plume merchants, hunters, and business lobbyists of the day underestimated Audubon's founding mothers. These were well educated, independent, and energetic women steeped in Boston's liberal social values whose educated elite nurtured one of the country's strongest abolitionist movements. (Mrs. Hemenway once opened her home to Booker T. Washington when he was refused accommodation at the city's hotels.) And not unlike some of the seemingly sudden societal awakenings of the 2020s, much of the country began to realize and grow disgusted with the wanton slaughter of wildlife and embraced the early environmental movement with surprising fervor.

In 1900, prodded by grassroots conservation advocates, Congress passed the Lacey Act (named for Iowa Congressman John Lacey), prohibiting the interstate trade in animals killed in violation of local state laws. As states passed increasingly strict bird protection laws and invested in agents to enforce them, the Lacey Act gradually put an end not only to the plume trade but also to market gunning, which had sealed the fate of

the Passenger Pigeon and the Eskimo Curlew and decimated the populations of many other wild bird species.

The bird conservation movement has from its incep-
tion been led by women who founded, supported, and
managed the advocacy, education, and habitat protection initiatives on which effective conservation must be
based. It is worth mentioning then that Harriet Hemenway, Minna Hall, and thousands of other women began
their movement 24 years before they were granted the
right to vote.

See also RSPB; AUDUBON.

Howard, Hildegarde (1901–1998)

Has been described as "the greatest avian paleontol-
ogist you've probably never heard of." At first a jour-
nalism major at UCLA, she was persuaded by a professor to switch to biology and, after taking courses in

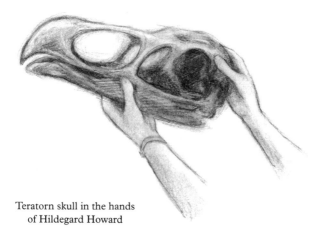

Teratorn skull in the hands
of Hildegard Howard

paleontology, eventually earned a PhD in that discipline at UC Berkeley with a dissertation on the fossil birds of the Emery Shellmounds in Emeryville, CA on San Francisco Bay.

Howard is best known for her discoveries at the La Brea Tar Pits in Los Angeles, including especially the so-called Rancho La Brea eagles and eagle-like vultures. She spent most of her career at the Los Angeles County Museum of Natural History, eventually becoming chief curator of science. While there, she described avian fossils of 3 families, 13 genera, 57 species, and 2 subspecies.

Identification

Well into the twentieth century, bird identification was something that a "field ornithologist" did in his study at the end of a day or when foul weather kept him indoors. Having spent the day exploring the woods and wetland borders, shotgun under arm, eager to live up to Dr. COUES's injunction to bag "all you can get," the first "birders" returned home, measured their specimens, noted the color of their "soft parts," plugged their mouths with cotton, and then set about preparing them as "bird skins." This done, if the collector was puzzled by any of his new acquisitions, he took the specimen in question in one hand, Coues's *Key to North American Birds* (or other reliable reference) in the other, and proceeded to identify his bird. Did the long-tailed flycatcher in hand have "three or four primaries emarginate; crown spot yellow in black cap" or merely "one primary emarginate; crown spot flaming in ashy cap"? If the former, it must be a Fork-tailed Flycatcher; if the latter, a Scissor-tailed.

Birds are still identified "in the hand" by ornithologists and bird banders using keys that systematize measurements or combinations of obscure plumage details. And there are still instances in which taking a specimen may be necessary (if not exactly desirable). However, with the exceptional quality of optical and photographic equipment now available, it is increasingly possible to perceive even minute details of plumage to identify birds along with their age and sex without having the bird "in the hand." It is also true that all masters of field identification have probably put in their time at a banding station or poring over museum trays to gain an accurate sense of what they are looking for. But before the 1920s, identifying birds—especially "obscure" species such as sparrows and "fall warblers"—on sight was seldom attempted. About that time Ludlow GRISCOM and a rapidly expanding group of his protégés began to prove that the great majority of bird species could be positively identified in life with the aid of a pair of "field glasses" and a comprehensive knowledge of field marks. This insight, aided by the appearance of illustrated field guides and ever improving optical equipment, gave rise to a new era in field ornithology now known (with fewer elitist overtones) as "birdwatching."

Birdwatching and its sportier variation, "birding," suggest many things to many people (see BIRDWATCHING), but identifying species is central to everyone's notion of the activity. Many now-obsessive birders can remember a "spark bird"—a Magnolia Warbler glimpsed at scout camp or a spoonbill noticed on a business trip to Florida—that unaccountably inflamed a dormant curiosity about nature and made them desperate to find a book that would tell them what they had seen. Once you have identified your first bird, it is difficult to stop wondering about this world of hitherto unnoticed beauty that has suddenly materialized or to resist the satisfaction derived from the possession of arcane knowledge and the ease with which herons and ducks, woodpeckers, and orioles yield their identities to your increasingly sophisticated scrutiny in the early months. But the hook of fascination is not firmly set until the reddest of Northern Cardinals, the most radiant of Yellow Warblers have become just the slightest bit boring and you find yourself scanning the sparrow pages of your field guide; or concluding that all sandpipers suddenly look intriguingly different rather than depressingly alike; or taking an interest in those tiny moving objects near the sea's horizon that some show-off is presuming to name. For the animus that drives the serious modern birdwatcher is not primarily a yearning for aesthetic thrills or scientific discovery; it is not even the "lure of the list" (though all these may be important ancillary rewards). Rather, it results from the evolutionary accident that many birds look a great deal like many other birds—and successfully parsing the differences combines the thrill of solving a difficult puzzle with achieving a kind of intimacy with other life-forms.

This truth soon appears in rudimentary form to the novice birder in the realization that mockingbirds and shrikes, for example, are superficially similar. But the paragon among today's birders is one who holds in mind every field mark, every call note, every nuance of gesture and posture (see JIZZ), every facet of distribution, seasonality, and extralimital occurrence of every species that could possibly paddle, flap, or hop into his line of vision. She knows where, when, and how to look for birds; he sees them and hears them with an acuity astounding to the uninitiated; they identify them at a distance or in motion confidently, rapidly, and (it usually proves) accurately. There is often an air of swagger in the manner of such paragons, but they readily and humbly admit their infrequent errors.

A unique pleasure of being a birdwatcher in the first decades of the phenomenon's existence was to watch the horizons of field identification pushed back. The legendary Griscom died puzzling over field problems now routinely solved by dedicated preadolescent birders. This casts no disgrace on Griscom's reputation; it is the logical result of an increased number of sharp eyes, ears, and brains scrutinizing birds in the field with the kind of systematic fervor that Griscom pioneered. One effect of the continual breaking of identification barriers is that rarities become less rare: Have the Eurasian stints suddenly become more frequent visitors to North America? Probably not, but there are certainly more people around who recognize one when they see it.

Another mark of advancing sophistication is the growing obsolescence of the standard field guides among birders of the upper echelons of birdernity. Members of this airy realm now take their instruction from journals of field ornithology and "postgraduate" field guides that unlock almost all the mysteries of unsongful *Empidonax* flycatchers and immature gulls and "peeps," and largely ignore "fall warblers," most of which have long since ceased to be confusing to them.

There is a grumpiness in some quarters about the alleged snobbery of so-called professional birders. "We," say the self-styled amateurs, "bird for fun," the implication being that the better you play the game, the less enjoyable it becomes. This is nonsense, of course, and the goal of total identifiability—however unattainable—is a healthy one, both intellectually and spiritually. If the hyperbirder is exceptionally liable to any character flaws, they are hubris (e.g., listing no unidentified alcids or jaegers for a day's sea watch) and humorlessness (e.g., discussing the possibility of an intergrade Kumlien's × Thayer's Gull with a straight face for more than a few seconds).

Those scanning this entry for the "how to" section will be disappointed, for, in the author's opinion, bird identification skills cannot be taught. There is undeniably a knack to the business, but there are no tricks that can be transferred verbally from one who knows them to one who does not in the way that, say, magicians can share expertise. Everyone must start from scratch, though, in fact, some begin before scratch by being born with keen eyesight and hearing, the most useful of birding faculties, which, unfortunately, are not dispensed equably or available on demand. Making the best of inherited traits, the would-be bird identifier should (1) Buy the best pair of binoculars and the best telescope affordable, learn how to use them as if they were essential prostheses, and keep them in the best condition. (2) Memorize—no, inhale—all available bird guides and all other literature relevant to identification (omitting books that purport to explain how to watch birds). (3) Spend every available minute of your life out looking for birds and examining them critically for their distinctive qualities. (4) Form an intimate friendship with someone as intelligent, as enthusiastic, and approximately as knowledgeable about bird identification as you are. If, after five years of following these guidelines faithfully, you are still saying things like "That Swamp Sparrow is much browner than the one in the book," you may have to resign yourself to remaining a novice birder for life. Happily, there is no shame in this; looking at birds seems to yield equal pleasure at all levels of expertise.

See also BIRDWATCHING; LISTING.

Intelligence

Terms such as "booby," "dodo," and "birdbrain" carry the unmistakable implication that birds are not very smart. And until about 20 years ago, science seemed to corroborate this assumption. The cerebral cortex—that great wrinkly mass of "gray matter" that dominates the appearance of the human brain and is known to be the source of our subtler tricks of ratiocination—is at best a smooth, thin, covering layer in birds. For early investigators, that pretty much settled the matter of avian intelligence. It has since been discovered, however, that a bird's "mind" originates in the relatively well-developed *corpus striatum*, on which the vestigial cortex rests. More specifically, avian intelligence has been located in a part of the *corpus striatum* called the *hyperstriatum* and learning centered in the bulge on the *hyperstriatum* called the *wulst*. In other words, in the evolution of intelligence, birds have taken an alternative anatomical path from that of mammals, with different parts of the brain developing to provide the physiological basis for intelligence. Another relatively recent insight is that while bird and mammal brains are both significantly more sophisticated than the brains of reptiles, there is great variation in the degree and nature of intelligence in these "smarter" classes. For example, members of some bird families do much better at certain mental tasks, such as counting and problem solving, than even relatively intelligent mammals such as monkeys. Other kinds of birds—such as pigeons, on which many early assumptions about bird intelligence were based—are dunces at the same tests. As avian IQs are more thoroughly investigated and compared, it is interesting to note that many of the birds that we intuitively think of as "smart," for example, members of the parrot and crow families, are indeed the whiz kids of the bird world.

Another factor that delayed the recognition that at least some birds are quite brainy was the once dominant influence of animal behaviorists who theorized that almost everything a bird does is "programmed," so to speak, into the genes it inherited from its parents and that its behavior throughout its life is essentially a combination of innate abilities (e.g., the ability to fly) and a series of unchanging, highly predictable responses to objects and events.

SO HOW SMART ARE THEY? The trap inherent in defining the intelligence of other animals is that we inevitably tend to use our own species as a kind of norm or, even more misleading, as the highest expression of evolution to date. As with the structure of the avian brain, it is more accurate to think of bird intelligence not as inferior or superior to the human version but as profoundly *different*. Birds experience the world very differently than people do—in many instances with superior equipment—and their needs, the driving force behind the adaptations of their brains, require different uses of intelligence.

One way of gaining an understanding of how bird intelligence functions is to observe how they use various well-defined forms of *learning*, the interface where instinct may be modified by experience. A very basic form of learning is known as *habituation*, in which young birds (and most other organisms) begin to distinguish shapes and actions that are harmless—for example, the movements of leaves—from those that are threatening. Many birds have learned that if approached nimbly and according to certain guidelines, fast-moving motor vehicles are not a threat and in fact can be a reliable source of food.

Imprinting is another fundamental form of learning. It has been shown that young birds begin to recognize familiar external sounds while still in the egg. Immediately after hatching, they tend to become strongly attached to the first moving object of a certain size that they see and hear. Normally, this is one of the parent birds, and the fixation initiates the chicks into a sense of identity with their own species and leads them into their characteristic life patterns as exemplified by their parents. But if, instead of first seeing their parents, young birds are immediately exposed to a different bird species, a person, or even a beach ball, their attachment to such unlikely "parents" can be no less ardent. If such chicks continue to be encouraged in their devotion to a beach ball and are deprived of any contradictory learning experiences, they may develop a permanent attachment and even, as they mature, direct sexual behavior to the stand-in, though this kind of prolonged attachment to the wrong parent is very rare. The initial gross association with one's own species in the first hours after hatching rapidly advances to more refined forms of discrimination with important implications such as choosing an appropriate mate and habitat selection in later life.

Another kind of perceptual learning involves *imitation and practice*. Experiments have shown, for example, that, though born with an innate knowledge of their song, at least some species never learn to sing it correctly unless they hear it "properly" sung and try to repeat it (see SONG). It is often difficult to tell where innate abilities leave off and learned skills begin. Watching young birds "learning" to fly, it is easy to assume that the process is a textbook example of "practice makes perfect." However, captive young pigeons have been denied this practice, as well as the opportunity to watch other birds, and when they were of the age when their siblings were flying skillfully, they were released and found to do just as well. In other words, the clumsiness of juvenile birds may have more to do with immaturity of muscle coordination than with lack of experience. In similar situations, parental "teaching" may consist mainly of stimulating young birds in an instinctual "following response," akin to the human propensity to yawn or vomit when others do. Whatever the origin, there can be little doubt that such experience increases proficiency to some degree.

Birds also learn by *trial and error*—the ability, for example, to recognize and avoid eating certain kinds of caterpillars that once (or repeatedly) made them violently ill. This can be seen simply as a form of conditioning, which "rewards" an animal for doing the "right" thing and "punishes" it for doing the "wrong" thing, thereby forming habitual patterns of behavior. However, Blue Jays can learn to avoid Monarch Butterflies by just watching other birds get sick, which properly belongs in the following category.

Insight learning is, so to speak, the ability to put 2 and 2 together without any previous experience with 4. It is in this realm of intelligence that bird performance is most impressive to people, since it approaches the deductive reasoning on which our own technologies and societies are based. And recent studies in this area have greatly enhanced our appreciation of bird brains.

Some of the most intriguing examples of insight learning in birds involve the use of "tools" to obtain food. Perhaps the best-publicized example of this is the practice among some African populations of Egyptian Vultures of breaking ostrich eggs by dropping rocks on them—documented and photographed for *National Geographic* in 1969 by the van Lawick-Goodalls. In another famous instance involving insight without tools, British Great Tits learned to pick open the caps of milk bottles to get at the cream at the top; this became so popular in Great Tit society that the milk company made stronger caps—which the tits soon mastered! Critics have made the case that these "innovations" arise by accident rather than intuition: the vulture, unable to pick up the huge ostrich egg in its bill, picks up a nearby rock instead in a redirection of its frustration; on rare occasions the rock is dropped on an egg, yielding a meal and conditioning the vulture to drop rocks on eggs. But this does not detract from the "innovators'" ability to grasp the significance of the event and repeat it—or, especially, from the "cultural transmission" of such bright ideas among whole populations of birds.

One of the most remarkable of breakthroughs in our knowledge of avian intelligence came from Irene Pepperberg's research on the Grey Parrot of Africa, especially an individual named Alex who now has his own foundation and website. Some of Alex's cognitive and communicative achievements are comparable to or exceed those documented for the brighter apes. He could, for example, with more than 80% accuracy identify and describe an object by a character not mentioned in the question: "Which object (of a varied assortment) is yellow, Alex?" Alex: "Key." He also grasped sophisticated concepts such as absence and same/different. He conversed and expressed desires spontaneously using an extensive English vocabulary. (Alex died in 2007, but his legacy continues through research with other Grey Parrots).

Irruption/Eruption

In an avian context, "irruption" denotes movement of large numbers of birds following the breeding season *into* areas beyond their normal range, as distinct from "eruption," which refers to movement *out of* the areas from which the birds originated. These are the orthodox terms for these linked phenomena, but such movements have also been called "invasions," "incursions," "flights," and "influxes." Partly because the species involved tend to erupt out of relatively remote regions, irruptions are typically more remarked by birders. Years in which irruptions of a particular species occur are commonly known as "flight years." Irruptions, as described below, are fairly well-defined phenomena and should not be confused with other movements and fluctuations in the numbers of certain species as a result of bad weather (Dovekie "wrecks," flocks of hurricane-driven Sooty Terns); intermittent range expansions (Carolina Wren); local breeding population increases related to food abundance rather than food scarcity (Black- and Yellow-billed Cuckoo populations tend to increase in areas where outbreaks of gypsy moths and/or tent caterpillars are occurring); or the characteristically erratic behavior of certain species, often tied to drought cycles (e.g., Dickcissel).

The families best represented among "irruptives" are the hawks, owls, jays, tits, and especially the cardueline finches (Fringillidae); however, individuals of other families, for example, Northern and Great Gray Shrikes, also qualify.

The characteristic these species share, aside from the irruptions themselves, is a fairly restricted food preference that is subject to sharp fluctuations—irregular or cyclical—in abundance. For example, the biology of many northern raptors is strongly influenced by the population cycles of the small rodents (e.g., lemmings and voles) with which they feed themselves and their young. Similarly, crossbills are dependent on cone crops in their native boreal forests. The "standard" model for irruptive behaviors holds that when food is abundant within a species' normal range, populations rise due to the ability to rear more young, and when this is followed by a corresponding contraction of the food supply, the now-augmented population must seek food sources elsewhere. This may indeed be the case for some species. However, recent studies of Snowy Owls have proven that while increased survival of young does indeed track with the booms and busts of rodent population cycles, it is actually during the years when food in the Arctic is most abundant that the young of the year appear in the greatest numbers to the south. Cone crops also follow cycles; for example, boreal trees follow two- or three-year masting rhythms, but these are greatly influenced by variables such as rainfall and temperature, and the irruptions of crossbills and other "winter finches" are therefore less predictable and can be much more dramatic.

Birdwatchers who live south of the ranges of irruptive species logically tend to perceive these movements as proceeding from north to south. But this is not always the case. Many birds of the boreal forest (e.g., Red Crossbill) are somewhat nomadic within that continentwide belt of conifers, so that there is often much east-west exploration before the depleted reserves of a whole region compel an advance southward. Similarly, in the last century, poor cone crops in the mountain pine forests of Mexico forced irruptions of Thick-billed Parrots northward into southeastern Arizona (though owing to habitat destruction on the birds' breeding grounds, they have not appeared north of the border in any numbers since the winter of 1917–1918).

The details of irruptive behavior vary from species to species. In some cases, the participants have been shown to be nearly all immature birds, which conforms with a tendency in birds of the year to wander farther afield than adults after they leave the breeding grounds; in other instances, adults predominate. In some instances, a significant percentage of some irruptive hordes may fail to find an adequate supply of food and may sometimes then die of starvation. Yet many individuals also survive to return to their "normal" area of residence. Crossbills and some other finches not infrequently will stay on well south of their usual breeding grounds to breed for a year or more. Irruptions may also be implicated in more lasting population shifts. Evening Grosbeaks were virtually unknown in New England and the Maritime Provinces of Canada until the winter of 1889–1890, during which the first of a series of irruptions from the West arrived in these regions. Though its abundance varies from winter to winter, the species is now regular at that season and nests regularly (if sporadically) throughout northeastern North America, though its abundance in the East generally has declined notably since the heyday of about 1955–1985.

Jizz (Not what you're thinking)

In birdwatching parlance, a distinctive physical "attitude" wholly apart from any specific field mark, which proficient birders can detect with consistent accuracy in identifying birds at distance or with only a fleeting glimpse. It can also be useful in distinguishing species of superficially similar appearance. The concept is essentially an amalgam of shape/structure, posture, and behavior. Knowledge of a species' jizz comes only after long experience with it and that of similar species in the field. In some cases, it is more reliable than rote observation of plumage details. Its use is crucial, for example, to the discrimination of diurnal raptors in flight, especially at a distance, and it is frequently helpful in sorting out shorebirds/waders.

The origin of the term has been much discussed among the birding elite. A plausible theory that has been widely offered is that it is a corruption of "GISS" (General Impression of Size and Shape) that was employed as a guideline for identifying types of aircraft by the RAF during World War II. But the term appears in British literature in the early 1920s, long pre-dating the coining of GISS. Another reasonable etymological guess was that jizz derived from the German word "gestalt," literally "form" or "shape" but deployed in psychology to refer to an entity that is more than the sum of its parts; but this idea has also been authoritatively debunked.

Kettle

A swarm of migrating hawks or other soaring birds spiraling upward on warm air currents. Because of the form of the rising "thermals," the birds riding them take on the shape of an inverted cone. This formation is known in American birding parlance as a "kettle" in reference to the whirlpool effect created when a pot of liquid is vigorously stirred. In northern New York, the term "boil" is used rather than "kettle," a reference to the circulation of bubbles in a boiling pot

of liquid. The Swedish word for the same phenomenon translates as a "screw."

See also SOARING (Thermal Soaring).

LBB/LBJ (also LGB, etc.)

"Little Brown Bird" or "Little Brown Job." Small birds of obscure markings and drab coloration that are therefore difficult to identify. For the beginner birdwatcher, most New World sparrows and female emberizid buntings are often spoken of in frustration as LBBs, while Old World warblers and small flycatchers in both hemispheres may be dismissed as Little Green (or Gray) Birds. The term was possibly coined during the tenure of Lyndon Baines Johnson, the late thirty-sixth president of the United States, who was often referred to by his initials. This gives "LBJ" a certain resonance among birders who were active in the 1960s and 1970s. But LBB now dominates.

Lear, Edward (1812–1888)

English author, musician, artist, and illustrator. Lear was an accomplished draftsman by age 16 with a particular interest in drawing birds. At 19 he published *Illustrations of the Psittacidae, or Parrots,* containing 42 hand-painted lithographs. He was apparently the first "bird artist" to work from live birds—which he sketched both in the London Zoo and in private collections—rather than from stuffed museum specimens (or, in Audubon's case, freshly killed specimens wired into particular poses).

Lear went on to provide illustrations for many other ornithological works and mentored Elizabeth Gould, who became an accomplished artist in her own right, providing many of the illustrations for the works of her ornithologist brother, John Gould, also an artist. And he became a noted landscape painter. But he is perhaps most famous for his nonsense verse, including limericks. Below are two examples that combine birds and rhyme.

> There was an old man with a beard,
> Who said "It is just as I feared!
> Two Owls and Hen,
> Four Larks and a Wren,
> Have all built their nests is my beard."

> There was an old man from Dumbree,
> Who taught little owls to drink tea.
> For he said "To eat mice
> Is not proper or nice."
> That amiable man of Dumbree.

Lek

Originally coined to describe the courtship arena of the Eurasian Black Grouse, the term now refers to any place where a number of male birds of a given species gather early in the breeding season to perform courtship displays with females in attendance. (It is also used for the display behavior of some mammals, including walrus and fruit bats, as well as some reptiles, amphibians, fishes, and insects). In North America, numbers of male sage-grouse and prairie-chickens gather on open "dancing grounds," where they affect highly stylized postures and gestures—spreading wing and tail features and inflating colorful air sacs on the neck—accompanied by a concert of arresting hooting sounds produced by forcing air in and out of their esophagus. In other grouse species, the cocks display alone but can be heard by birds on adjacent display grounds, an arrangement that has been termed an "exploded lek."

Great Bustard lek

LISTING 145

Lek behavior serves to establish hierarchy among males of a population and to facilitate female mate choice with the quality of the males' "costume" and performance, demonstrating his reproductive "fitness"; in at least some lek species, the cocks with the best moves get to mate with all the females (though some cheating is known to occur).

The Ruff, a largely Eurasian-breeding species of sandpiper, is another bird famous for lek displays, performed by males adorned in strikingly individualized expandable plumage "ruffs" around the neck. Some male Ruffs, known as "faeder" males, forego the fancy plumage, appearing instead in typical female plumage in spring, a ruse that permits them to mount females sneakily without having to compete directly with other males. Lek displays are also typical of bustards, as well as some hummingbird, parrot, and songbird species—notably the colorful manakins, cocks-of-the-rock, and umbrellabirds.

The word "lek" derives from the Swedish *leka*, which in certain contexts refers to "sex play."

Life Bird

A bird species that one has never seen before, and therefore may be added to one's "life list." See below.

Listing

The competitive branch of birdwatching in which one competes against oneself and/or others for the greatest number of species seen in a given place and/or time. Some listers reinforce their interest by keeping detailed records on phenological phenomena such as arrival and departure dates of migrant species near their homes or, among the more widely traveled, by recording faunal information for little-known localities. For most, however, sport and diversion are the chief attractions.

The American Birding Association (ABA) sets official standards for validity of records and provides a forum for discussion of listing issues in its flagship publication *Birding*. The ABA maintains rankings of state, provincial, ABA Area, and other species lists on their website.

The excerpt from *Birding*, reprinted below by permission of the ABA, articulates far better than any description the "lure of the list":

Bird-listing is a much-discussed topic in the birding world, especially among ABA members. Some birders malign listing, some can take it or leave it, some are disgusted by it, and others are compulsive about it. Because I find myself in the latter category, I would like to take this opportunity to tell others how, in my opinion, listing enhances the sport of birding.

First, I will list the lists that I keep. For a lister, listing lists is really fun, and it is something I seldom get a chance to do. Here they are:

1. U.S. Life List (526)
2. Minnesota Life List (341)
3. Month Lists (2,468; average, 206/month)
4. County Lists (11,924 for 87 counties; average, 137)
5. Season Lists (Spring, 312; Summer, 274; Fall, 292; Winter, 160)
6. Breeding-Bird List (235)
7. Yard List (3 residences) (highest, 135)
8. Early Spring Dates

LISTING 147

9. Average Spring Dates
10. Late Fall Dates
11. Average Fall Dates
12. Bell Ringers (birds seen in all months) (74)
13. January 1 List (98)
14. Day List (for each field trip) (highest, 177—May 22, 1976—Big Day)
15. Record Daily List by Month
16. Year List (since 1947) (highest, 296—1977)

Lists 2 through 16 refer to lists kept in the State of Minnesota, and Lists 1 through 13 are cumulative. All of these lists are self-explanatory, with the probable exception of 13. This is kind of a special list kept since 1949. It is a cumulative total of all species seen on January 1, the traditional day to take a field trip to start a new year list. The figures in parentheses are my personal species totals.*

Possibly, the first question many people will ask is, "Why keep so many lists?" My first and best answer is that list-keeping is fun because it gives you a chance to compare one year with the next, one day with another day, one area with another, and on, and on, ad infinitum; secondly, lists give me (and others) a good picture of the present distribution of birds in the state. I believe that there need be no other justification for lists. Probably the list that has become the most fun for me is the County List. My goal, to see at least 100 species in each county, has been accomplished in 82 of the 87 counties

---

* Author's Note: While impressive, this list of lists is not exhaustive. For example, Poop Lists (i.e., bird species spotted defecating) and Birds Heard from Bed are popular with some listers.

of Minnesota, with my average for the 87 counties being 137 species. The real fun of county listing is that it gets me into many different areas of the state, areas that probably would not be covered during regular birding trips. Some years ago, I became somewhat tired of always going to the same place at the same time of year. County listing provided the means to break the routine and get into new unbirded areas. To further enhance county listing, I started a town list. This is a list of all the named places in the state. To get to each of these places you have to cover the county "like a blanket." There are 1,828 named places in Minnesota—so far, I have been in all but 166 of them!

I have seen a number of birders become bored with seeing "the same old bird time after time." Although the thought of being bored by birds never occurred to me, one way in which I have escaped the possibility of boredom is by listing. Listing provides new challenges and an interest in seeing that "same old bird" in new areas and at different times of year.

—FROM ROBERT B. JANSSEN, "LISTING: MINNESOTA STYLE," *Birding* (AUGUST 1979)

Longevity

In birds, longevity can be determined with precision in two ways: (1) by recording the life spans of birds that are born and die in captivity, and (2) by banding birds of known age (e.g., nestlings) in the wild and recovering the bands upon their deaths. These two methods yield two different kinds of information on the ages of birds. The first tells us something of a bird's *potential* life span, that is, the greatest age a given species may attain under certain circumstances. The longest *reliably recorded* avian life is that of a male Sulphur-crested Cockatoo named Fred, hatched in 1914 and resident as of July 2020 at the Bonorong Wildlife Sanctuary, Brighton, Tasmania, Australia; Fred turned 106 in 2020. There are several probably accurate but less well-documented cases of cockatoos and macaws in the United Kingdom living to ages as high as 120 years.

Many other well-documented captive birds have had remarkably long lives. A Greater Flamingo in the Adelaide Zoo in Australia died at the age of (at least) 83 in 2010, and a male Pink (or Major Mitchell's) Cockatoo named Cookie lived at Chicago's Brookfield Zoo from 30 June 1933 through 27 August 2016, thus also attaining the age of 83. The *Guinness Book of Records* cites a Siberian Crane named Wolf that allegedly lived to 83 years in captivity. A male Andean Condor named Kuzya took up residence in the Moscow Zoo as an adult in 1892 and died 72 years later, thus probably about 80 years old, and a male of the same species named Thaao definitely reached age 80 at the Beardsley Zoo in Bridgeport, Connecticut, where he died in January 2010.

In general, bigger birds live longer than small birds, and birds that are known to have survived longer than 50 years include the hawks, eagles, condors, pelicans, flamingos, and parrots. The last are especially renowned for long life, perhaps because they are frequently kept as pets and their statistics recorded by doting owners.

Certain conditions of captivity that are not encountered in the wild—for example, lack of exercise, an artificial diet, air pollution in city zoos—may of course prevent a long life from being even longer, but in general, hazards to life in the wild are far greater and the chances of recording extreme natural ages are much smaller.

Banding records give good indications of *average* life spans as well as providing a few record ages under natural conditions. The albatrosses hold the present record for longevity in the wild. The Laysan Albatross Wisdom is the current record holder in this category. She was banded in 1956 as an adult on Midway Atoll, where she returns each year to breed, and is at least 70 years old as of this writing (2020).

It should be emphasized that the *natural* age limits of many birds that have proven to be long-lived in captivity are so far unknown, and new data will doubtless continue to break existing records. Conversely, the maximum age records for wild *songbirds* are probably close to extremes for old age. This highlights the distinction between longevity and *life expectancy*, which for most small birds is very short. From two to five years is probably close to the average life span for adult songbirds.

"Wisdom" the septuagenarian albatross, still raising chicks

Lumping (and splitting)

In an ornithological context, the practice of combining bird forms (especially subspecies) into larger taxonomic units (especially species). Taxonomists gain reputations as "splitters" or "lumpers" according to whether they perceive fine distinctions among closely related birds and classify them accordingly as full species (the splitters) or, contrariwise, merge two or more apparently distinct forms into one species (the lumpers).

Modern ornithology has been able to demonstrate that, in many cases, birds that seem superficially very different—for example, "Oregon" and "Slate-colored" Juncos—are, in fact, well-marked regional variations of the same species (in this case Dark-eyed Junco, *Junco hyemalis*), just as superficially very similar forms—for example, Bicknell's and Gray-cheeked Thrushes—have proven to be biologically distinct and thus merit recognition as full species.

For obvious reasons, instances of the former are not popular with those birdwatchers whose main pleasure comes from accumulating a long list of species (see LISTING). Some listers become so incensed by what they perceive as taxonomic larceny that they refuse to consider the scientific rationale for the decision.

Maturity

The age at which an organism is capable of reproduction. In birds, this varies from as little as 5 weeks (*Coturnix* quail) to 9 (or possibly 12) years in some albatross and eagle species. Most small songbirds breed for the first time when just under a year old, and many other species attain sexual maturity near the end of their second year. There is also some variation within species, some individuals maturing a year earlier than others.

Migration

Defined here as regular seasonal movements, as distinct from dispersal, migration is arguably the most spectacular and intriguing aspect of bird behavior. It involves about 80% of breeding species in temperate regions and may be observed over virtually the entire surface of the Earth from the high Arctic to the tropics and across all the oceans.

Almost everything we know about bird migration we have learned within the last 200 years, prior to which educated people could theorize that swallows spent part of the year hibernating in the mud or on the moon without fear of contradiction. However, while we still have much to learn, the literature on the subject is now so vast that only a few key elements can be briefly highlighted here.

WHY BIRDS MIGRATE. The dominant theories place periods of glaciation in the northern and southern extremities of the planet at the center of the mystery. As glacial advances made large areas of the Earth uninhabitable, birds were forced southward (and northward), toward the equator, into what is now the tropics. But as the glaciers retreated and the climate warmed, some species began to expand toward their previous ranges, though forced to retreat again each winter in order find their preferred foods of insects, fruit, and nectar. An alternative theory posits that many of our migrants evolved in the tropics but gradually learned that they could make a living farther north (or south, in the case of the Southern Hemisphere) as the climate warmed, but still had to commute back to their ancestral tropical home.

WHEN BIRDS MIGRATE. In general, birds move to and from breeding and wintering grounds in spring and fall, with peaks in April and May and September and October and with the corresponding times of lowest migratory activity in June and January. Arrival and departure schedules vary greatly among different species, depending on survival pressures. For example, many Arctic-nesting species must wait for a late spring thaw to ensure a reliable food supply and then depart before freeze-up, while songbirds farther south can arrive on their breeding grounds as soon as flowering plants have started attracting the insects on which they depend. One of the concerns about the effects of climate change on birds is that these evolutionary patterns will be disrupted, causing reduced reproductive success.

Different types of birds also migrate at different times of day. Raptors and other SOARING birds that require rising air currents move mainly from midmorning to late afternoon, and swallows and swifts, which feed on the wing, also travel by day. The majority of small songbirds, as well as many shorebirds, on the other hand, travel at night, while many waterfowl species seem to have no constraints regarding time of day. Needless to say, nocturnal migrants that meet the dawn over deserts or over water must carry on regardless of the hour.

DIRECTION AND ROUTES. It is widely and correctly understood that birds generally migrate on a north-south axis in both directions, though in many cases there is often variation such as northwest-southeast, depending on prevailing seasonal winds on which the migrants depend. Populations of species that nest on forest or tundra lakes in the center of a continent, but that winter at sea (e.g., loons and sea ducks such as scoters) may travel both due east and due west to reach the closest ocean.

While the public has been encouraged to believe that the majority of migrants follow a few major "flyways," in reality, daytime migrants follow so-called leading lines such as coastlines and river valleys that align with their prescribed direction, while nocturnal migrants travel over a broad front at altitude, guided by the stars (see NAVIGATION).

Many birds that summer at high altitudes move down to foothills or plains for the winter. Though they travel only some thousands of feet, these *altitudinal migrants* are responding to the same stimulus as the long-distance latitudinal migrants—namely, the need to retreat from a climate that cannot support them. This type of migration is very common, including in mountainous parts of Asia, South America, and Africa.

HOW HIGH AND HOW FAST? Nocturnal songbird and shorebird migrants normally travel at between 3,000 and 5,000 feet, but altitudes between 8,000 and 10,000 feet are not unusual, and radar scans have picked up flocks as high as 21,000 feet. Climbers summiting Mt. Everest have heard migrating Bar-headed Geese passing overhead at an estimated 28,000 feet.

Most migrating songbirds travel at airspeeds between 20 and 40 mph, larger species averaging faster than smaller species. Species of ducks, pelagic birds, shorebirds, and falcons seem to average between 40 and 60 mph, though small shorebirds flying at high altitudes have been clocked at airspeeds exceeding 100 mph. At least one Peregrine Falcon is on record at covering 1,350 miles within 24 hours. (For further detail, see ALTITUDE and SPEED.)

DESTINATIONS AND DISTANCE. As with other aspects of migration, variation is broad. Some birds travel only as far as necessary to accommodate basic needs, mainly food, so that some waterbirds will move only as far as the first unfrozen habitat they encounter, while many shorebirds are compelled by their evolutionary destiny to undertake maximal journeys from the high Arctic to the sub-Antarctic. A few of the documented avian distance fliers are as follows. *Arctic Tern* appears to be the record holder for the longest migration route, traveling 44,000 miles round-trip from high Arctic breeding grounds to Antarctic wintering grounds and back. Two seabirds that make a similar journey in reverse from southern extremities to northern are *Wilson's Storm-Petrel* and *Sooty Shearwater* (40,000 miles). The above three species are able to stop to rest at sea or land along their heroic routes, so perhaps at least as impressive is the *Bar-tailed Godwits* that fly 7,145 miles *nonstop* from Alaskan breeding grounds

Bar-tailed Godwit,
longest nonstop migrant

to winter quarters in New Zealand. Few landbird migrants approach these distances. However, *Northern Wheatears* that are circumpolar Arctic breeders travel across 9,000 miles of ocean (either Pacific or Atlantic) and desert to reach their wintering destinations in sub-Saharan Africa and Southeast Asia; *Swainson's Hawks* breed as far north as central Alaska and winter almost exclusively in the Argentine pampas; some *Barn Swallows* nest in the sub-Arctic and winter as far south as Tierra del Fuego; and *Blackpoll Warblers* nesting in boreal Canada arrive on the Atlantic coast in fall, double their weight by binge feeding, and then take off (typically following a cold front from the northwest) with enough "fuel" on them to fly up to 120 hours nonstop to wintering grounds in northern South America.

See also IRRUPTION; MORTALITY; NAVIGATION; RADAR.

Mobbing

The harassing of a potential predator, often by a mixed flock of smaller birds. Owls and hawks are frequently victims of noisy groups of tits, nuthatches, warblers, wrens, and other species, which call at, fly at, and, rarely, even strike at the presumed threat. Grackles, jays, and other nest robbers are often "mobbed" in flight by tyrant flycatchers and other medium-sized species, and flocks of crows and jays harry owls and hawks in a similar manner. Snakes and mammalian predators such as foxes may also be mobbed.

It seems clear that the activity is a collective response to a common danger (participants are known to cross territorial boundaries in joining such mobs), but whether

Songbirds vs. Eastern Screech-Owl

the intention is to drive the enemy away or simply to alert the avian community to the threat is unclear. The degree of feverish "displacement" activities (feeding, preening) often evident among members of a "mob" suggests that their response is an ambivalent mixture of fear and aggression. Scientists have opined that mobbing may function to help young birds learn to identify predators (the inability to recognize predators has led to the complete failure of reintroduction projects for species such as Thick-billed Parrot and Whooping Crane).

It is, of course, the mobbing response that birders take advantage of when they imitate alarm calls and owl "songs" in order to attract their quarry.

Mortality

Reminded though we have been recently that humanity is hardly sheltered from nature's deadliest ravages, we can still be appalled at the comparatively huge mortality rates regularly sustained by birds and other forms of animal life. It has been estimated, for example, that during the nestling stage, when vulnerability is generally greatest, the average mortality rate among passerine birds is about 50%. Within this average there is enormous variation among species (and populations) based on biology, food supply, weather conditions in a given season, and many other unpredictable factors obtaining only locally. But the average shows that for every circumstance in which 90% of passerine nestlings survive, there exists another in which 90% perish. Nor does maturity bring any great security. The toll that weather, predation, and the "success" of our own species take on bird populations makes a bird species' normal life span often totally inconsistent with its expected LONGEVITY in captivity, at least in species that thrive in captive settings.

The many instances of awesomely vast "natural" mortalities of birdlife now documented seem at first to encourage the notion that bird populations must be invulnerable to any threats posed by human agency. But natural mortalities are redressed by long-term counterbalances that have been built into the rhythm of nature over eons. Species that are subject to heavy predation or that suffer great annual losses due to the many hazards of long-distance migration, for example, tend to compensate for the drain on their overall population with such biological characteristics as large clutch size and number of broods, density within breeding niches, broad total range, colonial nesting, etc. Thus, when 148,000 waterfowl succumb to hailstorms in Alberta over a 22-month period, or untold thousands of songbird migrants are drowned in Lake Huron due to adverse weather conditions during their crossing, or like numbers of Dovekies are stranded ashore by fall gales and die of starvation or predation by gulls, the populations of the species involved recover fully and, as it were, seemingly automatically. We may notice temporary population declines locally, as in the well-documented decimation of the northeastern bluebird population during early spring ice storms in 1940 or declines of Carolina Wrens after very harsh winters in the Northeast (e.g., in 1978–1979), but we can confidently predict a rapid recovery from such natural cataclysms, unless, like the California Condor, the species as a whole is reaching the end of its tenure on the planet.

So integrated into the maintenance of population stability are such normal mortality factors as weather and predation that, viewed objectively, they can be seen as beneficial influences that keep us from being buried in warblers or Dovekies.

In sharp contrast to the ecological give-and-take that ultimately permits the harmonious coexistence of an unimaginable variety of organisms are the occasional great upsets that devastate a wide range of organisms on many fronts and thereby ultimately alter the whole system profoundly. One good example of this sort of upset is a major ice age. Another is humanity's recent domination of the Earth. The resiliency of bird populations is amply demonstrated in their ability to sustain multiple losses such as those cited above, but within the last 300 years or so they have ceased to be a match for our ubiquity and our still increasing efficiency in altering the planet's surface and atmosphere. For example, with one invention alone—firearms—coupled with the insatiable appetite of our burgeoning population, we accomplished in a few decades what previously was reserved for "acts of God," namely the extinction of entire species. Yet guns are now the least of our life-threatening inventions, and the ability of birds or anything else to "bounce back" from the cumulative effect of deforestation, oil spills, pesticides, radio towers, automobiles, acid rain, picture windows, domestic animals, thermonuclear weapons, etc., must be deeply doubted.

Having overwhelmed the natural balance, we rely on our alleged ability to balance the world ecosystem ourselves. Regarding the likelihood of this, the optimist points to our retreat from the wanton slaughter of egrets for hat decorations (see HEMENWAY, RSPB), the cynic to our retreat from environmental protection policy in the face of minor reductions in the gluttonous living standard to which we in the West have recently become accustomed.

See APOCALYPSE for more on some of these hazards; see also LONGEVITY.

Music

The songs of some of the most familiar European songbirds, especially the Common Cuckoo and Common Nightingale, have been incorporated now and then as themes in classical music. These species, plus the Common Quail (*Coturnix*), can be heard in the second movement of Beethoven's Pastoral Symphony. On the whole, however, bird "music" does not correspond very closely to the characteristic range of melodies and rhythms of Western music and therefore is seldom transposed successfully or at least recognizably. Other traditions, for example, Oriental flute music, seem to lend themselves much more naturally to the incorporation of birdsong.

The occurrence of bird themes in musical *lyrics* is, of course an altogether different matter. As with recording avian appearances in POETRY, noting instances of birds in human song comprehensively would require the lifetime services of a professional bibliographer. Suffice to say that bird lyrics may be found throughout all (or very nearly all) musical traditions, from the oldest folk songs to Gilbert and Sullivan's tit-willow—not to be confused with the Willow Tit (*Parus montanus*)—to the Beatles' "Blackbird" (*Turdus merula*) on the White Album.

The most famous "musical bird" is undoubtedly Mozart's European Starling (*Sturnus vulgaris*), a bird the composer purchased on May 27, 1784. A snatch of song sung by the pet starling may have inspired a passage in the composer's Piano Concerto in G major (K453), and, in any case, the bird could imitate the phrase so accurately (only substituting G sharp for G natural) that Mozart penciled in the starling's version under his own, adding the note: "Das war schön" ("that was beautiful").

See also VOCAL MIMICRY.

N avigation

Once science confirmed the astonishing fact that most of the birds of the North Temperate Zone travel hundreds or thousands of miles each year to more southerly wintering grounds (see MIGRATION), it was faced with the need to explain how that is possible. How does a bird with its apparently limited intelligence find its way to and from remote and restricted breeding and wintering areas? How does it know in which direction to go? How does it maintain its course, particularly in the dark, when many migrants travel? How can it tell when it has arrived at the right place in the Central American rainforests or Caribbean highlands? Does it really know where "home" is, or does it just follow an innate "flight plan" blindly? Does it know where to go "instinctively," or must it learn by experience—or both?

The sensory processes that encompass these difficult "hows" have so far been only partially explained. Some of what has been learned—almost entirely since 1940—is summarized below.

GENERAL SUMMARY OF PRESENT KNOWLEDGE. There is experimental evidence to support the following facts and speculations:

— *Spectacular "homing feats"* are performed by many species of birds, involving great distances and radical dislocations.

— *Navigational ability varies widely among different species of birds.* At least some species that typically spend their whole lives within a limited area have comparatively simple "homing" abilities, which would seem to correspond to their limited needs; species that undertake long migrations have evolved more sophisticated techniques, and nocturnal and pelagic migrants, which normally must find their way without the aid of landmarks, probably have the most highly developed orientation abilities of all.

— *The use of landmarks plays a significant role in homing for many birds.* Species of limited range memorize key features of their neighborhood and can be taught to learn landmarks of a wider area. Birds deliberately disoriented often wander at random, apparently searching for familiar places or "clues" such as a coastline that may bring them back into their familiar home range if followed. Birds that depend on the celestial clues described below may follow coasts or mountain ranges (leading lines) when the sky is overcast.

— *Birds have some form of "internal clock"* that allows them to compensate for the sun's normal daily movement and keep a steady course.

— *Nocturnal migrants depend partly on the stars for orientation.* Experiments with caged migrants showed that birds presented with artificial "skies" for a given migratory season oriented in the appropriate direction; when the sky was reversed the birds changed course accordingly, and when the sky was obscured they became disoriented. By selectively blocking out key stars such as the North Star and major constellations, scientists showed that Indigo Buntings could use any one of several celestial patterns if others were obscured.

— *The direction of migration (north or south) is influenced by hormonal predisposition* at least in some species. By artificially altering day length for different groups of caged birds, behavior ecologist Stephen T. Emlen was able to stimulate physiological "readiness" to migrate north in one group and south in another. When he exposed the birds to the same false sky, the birds followed their hormonal drives rather than the available star map.

— *Birds can sense and orient according to the Earth's magnetic field.* This has been shown in controlled experiments both by attaching magnets to birds, thereby disrupting the sensation produced by the Earth's magnetism, and by observing birds' reactions to an artificially created field similar in force to the Earth's. In the 1970s neurobiologist Charles Walcott and others located a small structure between the brain and the skull in pigeons that contains magnetite and that is used to sense the Earth's magnetic field for the purpose of orientation. There is evidence that a bird's magnetic compass may be a fundamental innate navigational system that can function in the absence of observed cues such as the stars.

— *Navigational ability can be acquired from experience* at least in some species. Experiments have proven that displaced adult European Starlings readily corrected course to reach their traditional wintering grounds, whereas inexperienced juvenile birds from the same population seemed to know the direction they should follow but could not adjust their course when they were displaced. The implication is that while they are born with some basic instructions about which direction to go and perhaps how far, juvenile birds of at least some species must learn the more detailed geography of their specific wintering areas.

— *At least some birds use smell to find their own nest sites* once they are, so to speak, in the neighborhood. This has been experimentally demonstrated in Leach's Storm-Petrels, which fly downwind of their colony on arrival to pick up and follow the scent of their particular nest burrow. This would seem to be especially practical for birds such as many seabirds that must find a tiny burrow among hundreds at night amid acres of similar habitat; these birds are known to have an acute sense of smell, which they use to locate food as well as home.

— *Migration routes of species and populations may change* more often than we have previously supposed, and the genetic programs that set a bird's course can apparently evolve rapidly to accommodate the new travel plan. In recent years some Eurasian Blackcaps (a species of Old World warbler) from Germany began wintering in the British Isles rather than their "normal" wintering grounds in the western Mediterranean and North Africa. The young of the British-wintering Blackcaps showed an innate tendency to head northwestward rather than southwestward in the fall.

See also MIGRATION.

Nests and Nesting

Asked to picture a bird's nest, most people would probably imagine a cup made of grasses or other vegetation. This is not an inaccurate description of many songbird nests but is woefully incomplete in describing the nests of the world avifauna. What follows is an attempt to paint a more detailed picture.

WHY MAKE A NEST? This in effect is what a number of species in different families seem to have asked themselves, since they make no nest at all, simply depositing their eggs onto a cliffside shelf (auks, falcons); in a shallow "scrape" on a beach (plovers, terns); on top of some leaf litter in a shrubby opening (grouse, nightjars); or in the abandoned nest of another species (owls, some ducks). In some cases, a bit of grass may be added to these spartan accommodations by way of a lining. For protection from predators (or intrusive birdwatchers), these avian minimalists rely on aggressive site defense or the cryptic coloration of the sitting parents that matches the ground cover.

While a number of no-nesters have made a success of their low maintenance approach, it also seems clear from their elaborate evolution and general prevalence among most birds that nests have aided survival by protecting eggs and young from weather and depredation and perhaps by increasing the efficiency of the rearing process.

ARCHITECTURE. The "styles" of nests are wonderfully varied among different groups of birds. Start with the "standard" cup nest of most songbirds which by way of variety may be sited on a wide branch, in a crotch, or on a level surface on a building; variations include the vireos' tidy woven cups bound with spider silk and suspended from a small tree fork and the pendulous sacks woven by some oriole species. Many wren species build rather large, loose, domed nests of sticks or grasses in trees, shrubs, cacti, or cattails—or make do with a rock cavity, rodent hole, bird box, or more unusual venues (Audubon depicted a House Wren family nesting in a discarded hat!). Stick constructions—from the flimsy little collection of twigs on which many doves place their eggs to the rather sturdier platform of branches favored by herons and other wading birds and the massive wood piles accumulated year after year by large raptor species (see below)—are another major type of egg support. Then there are the tree cavities excavated by the world's 234+ species of woodpeckers, whose holes vary in proportion to the size of the carpenters (3.5 inches to almost 2 feet in diameter) and which are "sublet" to a wide range of other birds from tits to owls; and the earthen tunnels and chambers excavated by kingfishers and their relatives and some swallows. Among the most aesthetically pleasing of nests are the clustered, flask-shaped "adobe" homes of Cliff Swallows and House Martins.

A few nesting habits deserve special mention. A single family of 12 species of chicken relatives, the Australasian megapodes, bury their eggs in earth warmed by the sun or by volcanic activity or under decaying vegetation, thereby relieving themselves of the duties of incubation (though trading these for the perhaps only slightly less tedious "oven-tending"). Grebes and a few other species of birds cover their eggs with dead plant material when they leave their nests temporarily, and it has been suggested that this, too, is a form of "carefree" incubation.

CONSTRUCTION MATERIALS commonly used for basic external nest structure include sticks, grasses, cattail, sedges (*Scirpus*, etc.), rushes (*Juncus*), seaweed, wet decaying aquatic plants, Spanish moss (*Tillandsia usneoides*), bark, foliose lichens, paper, string, "debris," mud, and guano. Some birds typically "decorate" their nests' exteriors with flowers or colorful bits of yarn (or plastic), and most Great Crested Flycatcher nests incorporate a length of shed snake skin (!).

All passerines and many other species line their nests as the last stage in the construction of complex nests, and the materials used are much finer than those used in the "walls." Typical lining materials are leaves (dead or fresh), fine grasses, *Usnea* lichens, fungal fibers (mycelia), mosses, plant down (thistle, milkweed, etc.), bark fibers, pine needles, contour feathers, down feathers (ducks and geese only), and animal hair/fur.

Many species solidify their nests with some form of binding material. This process varies widely in both extent and materials used. Some swallow nests are made exclusively of mud. The New World phoebes, which place their nests on rock ledges or man-made structures, build largely of mud but with a larger proportion of plant material. Many thrushes "plaster" their interior walls with mud and then add a further lining of fine grasses. Crows sometimes use small amounts of mud or animal dung. And the excrement of nestlings adds to the solidity of the nest structure in some species. Swifts (most species) are unique in cementing their cup nests of twigs with their own gluey saliva, which forms almost the entire nest in some Asian swiftlets. And hummingbirds, kinglets, and other small species use spider web and caterpillar silk to bind the fine materials used in their diminutive nests.

SIZE. It will surprise no one that small birds build smaller nests than large ones. The span however is impressive: the tiny Calliope Hummingbird's minute cup measures $1\frac{1}{2}$ inches in diameter by $\frac{7}{8}$ of an inch high and weighs perhaps an ounce, whereas a Bald Eagle's nest accumulated over decades may attain a diameter of 9 feet, a depth of 20 feet, and eventually weigh a ton or two.

LOCATION, LOCATION, LOCATION. Birds' nests are found in virtually every conceivable situation except in midair, on the surface of the sea (the legends of the HALCYON to the contrary notwithstanding), and underwater (though the nests of grebes are typically partially submerged). Placement ranges from 3 or more feet below the Earth's surface (Burrowing Owl, some auks)

to more than 100 feet in trees (kinglets, some woodpeckers); the record holder in this category is the Mar-
bled Murrelet, a small Pacific seabird, whose nesting
wasn't known until August 7, 1974, when its nest—a
small cup, made mainly of guano—was discovered ac-
cidentally by a tree climber 148 feet up in a Douglas Fir
in Big Basin State Park, Santa Cruz County, California.

Particular species tend to nest within a characteristic
range of heights, which, however, may be quite broad

and span two or more niches. Ground nesters never nest in treetops and treetop species do not nest on the ground, but some ground nesters will also build in low shrubs or the lower branches of trees, and treetop species may descend to within a few feet of the ground. This variation may be seasonal and geographical as well as individual. Many species have come to regard man-made structures as acceptable nest sites, and a few—Barn Owl, some swift species, Barn and other Swallows, phoebes, House Sparrow, House Finch, Chimney Swift—have come to prefer them to what nature offers. Barn Swallows, House Sparrows, and wagtails have been recorded with some frequency nesting on ferries, following the craft on its daily rounds to brood their eggs and feed their young. Purple Martins in eastern North America appear to nest *solely* in "martin houses" and hollow gourds set out specifically for them.

Some birds enhance nesting security—whether by design or by chance—by nesting close to other, more aggressive bird species. For example, several species of songbirds have occasionally nested among the twiggy interstices of a raptor's nest, and birds nesting within tern or gull colonies doubtless derive some benefit from the latter's no-nonsense approach to nest defense. Female Black-chinned Hummingbirds often site their nests near nests of Cooper's Hawks and Northern Goshawks, to discourage Mexican Jays and other nest predators from depredating the nest. A number of Neotropical bird species routinely nest near wasp or ant colonies, thereby deriving protection from predators without apparently incurring any risk to themselves; this technique is apparently not documented as yet for any North American species.

WHO DOES WHAT? Site selection and nest building tend, on the whole, to be dominated by the female. On the other hand, males are typically not exempted entirely from these duties. Ornithologists have determined that in 56% of North American passerines, males made some contribution to nest building. In most species they play only a minor or ritual role, such as bringing nesting material to the female, but in others (e.g., gnatcatchers), males are very effective nest builders. The Red Phalarope may be the only species in which the male alone is responsible for nest construction, although unmated male cormorants are known to build a nest, which is then rearranged by the female if and when she consents to mate. Among *all* hummingbirds; *most* tits, icterids, tanagers, and finches; and *some* flycatchers, swallows, vireos, and wood warblers, the female alone builds the nest. In most species, nest building is shared in a variety of ways. For example, male wrens of some species build a series of nest "shells," and the female selects one and lines it for occupation; however, in House Wrens, the female does all the real work, with the male reduced to ritualistic stick carrying.

See also EGG.

Nice, Margaret Morse (1883–1974)

In the bird world, the name Margaret Nice is inextricably associated with the Song Sparrow, due to her meticulously detailed *Studies on the Life History of the Song Sparrow* (1937). Nice's interest in birds began in childhood in Amherst, Massachusetts, where she started taking notes on bird behavior starting at age 12. After college (B.A. from Mt. Holyoke; M.A. in biology from

Clark), marriage to a professor of medicine, and the birth of five daughters, she moved with her family to Oklahoma (1913–1927), where her husband had accepted a professorship at the University in Norman. Here she continued her bird studies, eventually publishing *Birds of Oklahoma* in 1931. Her avid intellectual curiosity and watching her children grow up also drew her to child psychology, especially language development, on which she published 18 papers.

Another academic move to Columbus, Ohio, allowed her to connect with a community of professional ornithologists, including one of the only other prominent female ornithologists of the time, Florence Merriam BAILEY (who is reputed to have addressed Margaret at an American Ornithologists' Union meeting as "Mrs. Mourning Dove Nice," referring to her admired studies of that species). It was here that she began her seminal work on Song Sparrows, devoting eight years to the study of more than 70 banded pairs of the species. The publication of this work brought her international recognition and contact with some of the most respected ornithologists of the era, not only for the quality of the work, but also because it was an early (and welcome) signal of a shift from the obsession with discovering new species and distributional studies to a behavioral focus. The eminent Ernst Mayr allowed that she had "almost singlehandedly initiated a new era in American Ornithology."

Nice published more than 200 papers on birds and 3,000 book reviews as well as a popular account of her Song Sparrow work, *The Watcher at the Nest* (1939). She was inducted into many ornithological societies worldwide, received an honorary doctorate from Mt. Holyoke, and was only the second woman to receive the prestigious Brewster Medal, after Florence Bailey.

Nuttall, Thomas (1786–1859)

Born in Yorkshire, England, where he worked as a printer's apprentice until his emigration from England in 1808 to Philadelphia, then the "Athens" of American natural history, Nuttall was chiefly a botanist. He made extensive collecting expeditions alone and on foot across much of North America, especially in the South and West. He took an ill-paid curatorial chair at Harvard for 11 years—a period in his life that he described as "vegetating with vegetation"—and accompanied the Philadelphia Academy of Natural Science Columbia River Expedition headed by John K. Townsend. In 1842 he inherited an uncle's estate in England and returned there to live for the rest of his life. Nuttall's greatest works were botanical; however, he also produced an account of some of his travels, called *Journal of a Journey into the Arkansas Territory* (1821), and two small volumes that might qualify as the first field guide to North American birds. Despite his botanical bent, his name is memorialized in that of the Nuttall Ornithological Club, the oldest of its kind in North America (founded in 1873) and the progenitor of the American Ornithologists' Union (now the American Ornithological Society). The Common Poorwill (*Phalaenoptilus nuttallii*) was named for him by Townsend, who also named a monotypic genus of tyrant flycatchers for him. But *Nuttallornis* ("Nuttallbird"), which once contained only the Olive-sided Flycatcher, has now been subsumed in *Contopus*, home of the pewees.

Odor

Aviculturist: I plugged my parrot's nares with cotton yesterday.

Ornithologist: Really? How does he smell?

Aviculturist: Terrible!

Not surprisingly, the body odors of birds have so far proven of little general (or even specialized) interest, and literature on the subject is scant. Nevertheless a few broad comments may be ventured if for no other reason than to encourage pioneers into this ornithological frontier. It is probably obvious that bird colonies or nest sites, where excrement and putrefying food matter are allowed to accumulate, may have a very pronounced odor that is offensive to humans in many cases. However, birds themselves may have characteristic odors, ranging from stenches to fragrances. Vultures, some storks, and other carrion feeders tend to carry the unpleasant (to us at least) smell of their preferred food. Great Horned Owls, which feed frequently on skunks, tend to acquire more than a hint of butyl mercaptan, the technical name for skunk stink, which can even overpower the naphthalene with which museum trays of bird specimens are treated.

The Hoatzin, a unique species inhabiting swamps in the tropical lowlands of South America, is known as the "stink bird" because it gives off a strong odor of fresh cow manure; the smell results from the bird's diet consisting of 82% leaves which it digests by means of bacterial fermentation in its greatly enlarged crop, much as cows and other cud chewers do in their rumen. Petrels and other tube-nosed seabirds have a distinctive oily or musky odor, a mild version of the strong-smelling stomach oil they are liable to vomit onto molesters of

Hoatzin, a.k.a. stinkbird

any and all species. Most land birds are reputed to have little or no natural odor; however, to this writer's nose, at least some shorebirds and songbirds have a pleasant, almost floral (though not sweet) smell. Doug Pratt, an authority on Hawaiian birds, has pointed out that Hawaiian honeycreepers have a very distinctive odor that survives death and smelly preservatives, and he has offered the *lack* of this aroma as one of his reasons for *excluding* the Poouli and members of the genus *Paroreomyza* from the honeycreeper family.

Bird banders/ringers notice various smells of our polluted environment (e.g., sulfur dioxide) in the plumage of nocturnal migrants, which have presumably arrived in their mist-nets via city skies.

As a final proof that much work remains to be done in this field, I cite the distinguished ornithologist Robert Storer, who has affirmed that Crested Auklets "give off an odor like that of tangerine oranges." (!)

See also SMELL.

Origins of Birdlife

Despite the difficulties involved in tracing avian lineages, some highly educated guessers have speculated very convincingly about the origins of some bird families. Ornithologists Ernst Mayr and Lester Short thought it impossible on the evidence available to postulate the origins of 29 families worldwide, including most waterbirds and some families of cosmopolitan distribution, such as hawks, nightjars, and swallows. Of the bird families that currently occur in North America, they believed that the American vultures, grouse,* turkeys, limpkins, wrens,* mimic thrushes, waxwings,* silky-flycatchers (e.g., Phainopepla), vireos, and wood warblers evolved in North America (including Mesoamerica and the Antilles in Mayr's definition); with recent insights in bird taxonomy, the gnatcatchers, American cuckoos, and American quail have been added. Families with asterisks are the only ones that have reached the Old World—one species each in the case of the wrens and waxwings. The shallow sea bed of the Bering Strait that periodically bridges the watery gap between what are now Alaska and Siberia during times of low sea level is believed to have been a major route for interchange of New and Old World avifaunas. Excepting the grouse, all the families listed above now occur in the Neotropical region as well.

Families believed to have originated in South America are the hummingbirds, guans, tyrant flycatchers, true tanagers, and American blackbirds/orioles. None of these families occurs in the Old World; all reach greatest diversity in the Neotropics.

The pheasants and Old World quail, cranes,* pigeons and doves, Old World cuckoos, barn owls, typical owls, kingfishers, larks,* crows and jays, tits,† nuthatches,† treecreepers,† thrushes, Old World warblers,* kinglets,† pipits and wagtails,* and shrikes* originated in the Old World, according to Mayr. Those marked with asterisks do not reach the Neotropics or only barely so; daggers mark families that reach only the highlands of Mexico and Central America. Since Mayr and Short's time, there has been a profusion of papers theorizing on the geographic origins of modern bird families, but many of their early suppositions still stand.

Ornithichnite
The fossilized footprint of a bird or birdlike dinosaur.

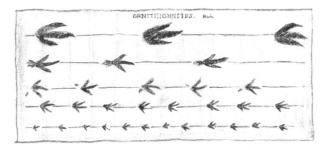

After a drawing on linen by Orra White Hitchcock, ca. 1840

Ornithomancy

Magical prediction of the future by observing the behavior of birds. Although ornithomancy would seem to be a practice of the distant past, its traces haunt our language to this day. In ancient Rome, an *augur* was an official diviner who interpreted omens before public events, usually based on behavior of birds or other animals, or sometimes on celestial phenomena. The word probably derives from the Latin *avis* (bird) and *garrire* (talk), and we still say that something "augurs well" (or poorly) for a certain outcome. The Romans also used the term *auspex* (from *avis* + *specere*, "to look") for the position of an *augur*. Our English words "auspices" and "auspicious" thus also descend directly from magical Roman birdwatching.

Painting

There are two discernible kinds of "bird painting": (1) works of art by acknowledged masters in which birds figure—sometimes very prominently—in the composition; and (2) bird portraiture in which birds are treated for their own sake as the main objective of a painting. (The Swedish impressionist Bruno Liljefors may be the only painter who managed to be both a brilliant *bird* portraitist and an acknowledged mainstream artist.) The best examples of the latter may exhibit exquisite draftsmanship together with fine perceptions of light, texture, composition, and other values. But their *intent* is altogether different from that of the former, and so this form of bird painting is treated separately below.

ART HISTORY. There is little exaggeration, if any, in the statement that you can find birds wherever you look in the history of art. They occur most often as genre elements, included, so to speak, to show the world as it is (or was). Currently, the oldest known bird images in which the species is recognizable are red ochre portraits on cave walls of a giant, flightless extinct species of *mirihung*, or "thunder bird" (*Geryornis newtoni*), a massive relative of the Emu, believed to have been painted by Australia's earliest human inhabitants as long as 40,000 years ago. (The time of its extinction appears to coincide with the arrival of humans in Australia.) Paleolithic cave paintings of hunting scenes in Europe from about 17,000 years ago contain some identifiable mammals, but only very small anonymous bird figures, perhaps indicating their relative unimportance as game.

By contrast, herons and kingfishers are clearly portrayed in the limestone relief of a hippopotamus hunt in the tomb of Ti (ca. 2400 BC) of the Old Kingdom of Egypt. And one of the treasures of Tutankhamen's tomb (ca. 1360 BC) is a painted chest showing the young king hunting bustards (among many other animals) and attended by two Griffon Vultures. Hunting scenes with birds also show up in classical Etruscan, Greek, and Roman wall paintings and in Medieval tapestries and illuminated texts (falconry was a very popular genre subject in the Middle Ages), and continue to appear in serious works of art until nearly the present day (e.g., Winslow Homer's *Right and Left*, showing a pair of Common Goldeneyes being dispatched from a distant boat on a wintry sea). And the hunt continues to be a central theme in the art of African, Australian, South American, and Inuit cultures.

In the fifteenth century, common dooryard birds such as crows, magpies, and sparrows became prominent in

After Winslow Homer's "Right and Left"

carefully drawn scenes of daily life such as the Lim-
bourgs' *Les Très Riche Heures du Duc de Berry,* and later
in the town-and-country-scapes of Pieter Bruegel the
Elder (1525–1569). Though recognizable species occasionally appear in later landscapes, e.g., Rubens' *Landscape with the Chateau de Steen* (distant magpies), birds
of this period are more likely to be used as vague evo-
cations of nature, for example, the anonymous avian
shapes in flight, added with a few strokes of the brush by
Claude, Constable, and the Impressionists. The American Luminist Movement (1850–1875) made liberal use
in its seascapes of the kind of little wide-V "seagull" that
all school children learn to make and stick picturesquely
in their skies. A grisly variation on these obligatory seabirds depicts seabirds (gulls or petrels?) hovering over
the carnage in Turner's *The Slave Ship.*

The purely decorative qualities of birds also appealed to some of the earliest known artists, for example, Greek seventh-century BC perfume vases in the shape of an owl, or motifs on painted black-figure Attic kylixes (sixth century BC). The highly formalized garden scenes in late fourteenth- and early fifteenth-century tapestries and frescoes made good use of the forms and bright colors of such birds as peacocks and Eurasian Golden Orioles. And these "ideal" birds were also popular in sensual Arcadian scenes in the Renaissance (e.g., Titian's *Bacchanal*, 1518) and in the romantic eighteenth-century nymph-scapes of Watteau and Fragonard. On Baroque ceiling frescoes, cupids and angels usually outnumber birds, but not infrequently, a flight of swallows or some other suitably idyllic avian phenomenon is included, as in Guercino's ceiling fresco *Aurora* (1620s) in the Villa Ludovisi in Rome. Somewhat ironically, some of the most detailed depictions of birds in art are found in Dutch still lifes of the seventeenth century. As with the sumptuous flower and food arrangements of the same Baroque tradition, birds of many species, carefully chosen for the variety of their colors and patterns, are shown to exquisite advantage—except that they are all dead.

A few painters have recognized the potential of birds in enhancing fantastic themes. The unrivaled master of feathered fantasy is the northern Netherlandish master Hieronymus Bosch (1450–1516), who, to judge by the precision with which he painted actual species and the variety of birds he includes, seems to have been more than a little interested in birdlife. Birds appear in many of his paintings as elements of genre (see above), but in his famous *Garden of Earthly Delights* triptych he uses a staggering diversity of birds—real and imaginary—in three of art's most fantastic landscapes. In the Eden and Hell side panels, he gives us fairly standard idyllic (egrets, peacocks) and horrific (owls and nightjars) avian images, respectively. But in the center panel, showing either an Adamite heaven on Earth or (more probably?) "man in sin," the birds are far more than peripheral decorations. Giant goldfinches, woodpeckers, hoopoes, jays, and owls wade about in crowded pools with the same spirit of frolicsome liberation that their human brethren display; a smiling duck drops fruits into a man's mouth, spoonbills squawk in delight at riding on a goat's back, a European Robin is entirely content to carry a man with a huge seed pod on his head. To a modern eye, this scene (though somewhat overcrowded) might be seen as representing some eco-paradise in which birds and people (and a great many other things) seem to be living together in perfect contentment. But most art historians agree that, despite the innocent expressions of the creatures in this garden of delights, they represent the consequences of the Fall depicted in the left-hand panel of the triptych and the precursor to the eternal damnation shown to the right. Like virtually every other animal and shape in the painting, the birds all have an allegorical meaning relating to one or more specific vices that a contemporary viewer would have recognized.

Before leaving fantasy, we should probably give a respectful nod to the twentieth-century fantast Henri Rousseau, whose haunting jungle scenes (e.g., *The Dream*) often include a bird or two of imaginary species that help to evoke an aura of primitive innocence.

Finally, it should be emphasized that, especially in early painting, birds are often symbols for concrete entities, for example, saints or Satan, or for emotions or phenomena, for example, the volatility of the spirit or Spring. Bosch's owls, for example, stand specifically for evil/the underworld, omnipresent but often unnoticed in the human sphere; and his eggs, which bear surprising contents in many of his works, are alchemistic devices. Birds in early ecclesiastic paintings are also included in many cases to represent a specific aspect of Christianity, for example, the bleeding pelican, which appears in painted crucifixions (see RELIGION). A less literal form of symbolism is exemplified in Brancusi's gleaming abstract bronze, *Bird in Space* (1919), which embodies the sleek grace that man admires in the flight of birds.

Arguably the most exquisite of painted birds are those done in the tradition of the Northern Sung Dynasty painter-emperor Hui Tsung of eleventh-century China. The graceful egrets among lotuses and fierce goshawks perched in flowering plum trees also stand for qualities that they suggest to human observers, though no one can deny their decorative appeal. The elegant lines and harmonious forms are meant to express Buddhist teachings, such as the transmigration of the soul and the insubstantiality of the ego, as well as simply delighting the eye.

BIRD PAINTING. Defined here as the depiction of birdlife for its own sake—in an attempt to distinguish ornithological painting from birds that appear for a variety of reasons in works of the mainstream of the painting tradition (see above) and also from depictions of mythical birds, for example, the Thunderbird, depicted for eons by Native Americans. The unfortunate but inevitable implication of this definition is that paintings of birds (i.e., *about* birds) can never aspire to be more than minor works of art.

Leaving aside "What is art?" and "Why won't curators hang Fuertes next to Matisse?" and other vexed questions, it may yet be possible in the limited space afforded here to suggest an approach to evaluating the qualities to be found in "bird art," a genre that has never been more popular. We might begin by examining the abilities and probable intentions of the artist-craftsman. Some people learn to draw birds out of devotion to their subject matter, knowing and caring little about "art" and acquiring only as much knowledge of light, composition, draftsmanship, and atmospherics as is necessary to produce a comment such as: "Now *that's* a Blue Jay!" or "Why, you can see every feather!" Many such "bird artists" produce admirable—often much prized—work, which, however, does not aspire to be more than a "lifelike" and decorative depiction of an attractive animal. Others are artists first—schooled in the traditions and techniques of painting—but also have a deep understanding of the natural world and are committed to using their gifts and experience to express their acute perceptions.

The line just drawn is not a sharp one. Any trained artist who sets out to make a living painting birds will invariably spend at least his or her early years doing illustrations, which, because of their purpose (field identification, scientific description), allow for little interpretation. Such work cannot be expected to show an artist's depth or range no matter how fine the artist's sensibilities or sophisticated the technique. Contrariwise, the most talented of bird portraitists sometimes produce works of some depth and subtlety out of fortuitous combining of raw talent and their feeling for their subject. As a general guideline, however, we can expect even the most straightforward field sketches of the artist-who-paints-birds to contain a difficult-to-define quality (artistic imagination, aesthetic insight?) that is usually lacking in the grandest, most colorful designs of the bird painter.

Sadly, the public, and particularly the American public, is severely handicapped in its ability to develop any subtle discernment regarding bird painting. The best bird guides and the most magnificent of illustrated monographs can show only one aspect of an artist's ability and very little of his or her "vision." The sentimental animal prints now hawked in airports and in nature magazines at exorbitant prices do little to educate the public taste. And the most expressive works of our best natural history artists are mostly sealed from our view in the homes of the few who can afford to buy them as originals, because, justly or not, few art museums will hang a painting of a bird even alongside a minor artist in a mainstream tradition. So what we tend to see in infinite succession are bird portraits, some very fine, many incompetent or excruciatingly dull, but all of them too limited in concept truly to measure up to the term "bird art." Fortunately, a few fine arts books have begun to be published in this realm that illuminate the European tradition of the artist of nature. Preeminent in this tradition is the Swedish impressionist Bruno

Liljefors (1860–1939), who has influenced virtually every would-be painter of nature that followed. After decades of near obscurity in North America, his work can finally be seen (though, alas, poorly reproduced in many cases) in Martha Hill's *The Peerless Eye* (1987). Aficionados of this genre should also seek out the published work of Liljefors' most brilliant modern heir, Lars Jonsson, who has managed to be both an unsurpassed field guide illustrator as well as gallery artist of international stature; see especially *Birds and Light* (2002), which provides a historical overview as well as many examples of the artist's work. The work of Robert Verity Clem also belongs in this lofty realm; sadly he refused to publish further work after he was disappointed by the reproduction of his work in *The Shorebirds of North America* (1967), though in his later years he produced landscapes (with birds) that compare favorably with those of Andrew Wyeth. Any list of Master Nature Painters must also include John James AUDUBON; Louis Agassiz Fuertes (widely deemed to be Audubon's heir as the dean of bird portraitists); Allan Brooks, William T. Cooper, Don Eckleberry, Eric Ennion, Elizabeth and John Gould, Francis Lee Jacques, J. Fenwick Lansdowne, Edward LEAR, George Lodge, Roger Tory PETERSON, Peter Scott, Arthur Singer, George Miksch Sutton, Archibald Thorburn, C. F. Tunnicliffe, and Walter Weber.

For excellent surveys of current natural history art by living artists, see catalogues from the Lee Yawkey Woodson Art Museum in Wausau, Wisconsin (including sculpture and other media as well as painting), and the series of books published in the Netherlands by the Artists for Nature Foundation.

Those with a budding interest in this subject should also be aware that in addition to the Woodson Museum, two other North American museums specialize in art of the natural world: the National Museum of Wildlife Art, Jackson Hole, Wyoming; and the Massachusetts Audubon Society's Museum of American Bird Art, Canton, Massachusetts.

Peterson, Roger Tory (1908–1996)

With the possible exception of John James AUDUBON, the man most closely associated in the public's mind with birds and, especially, their identification in the field. As with many life-long birdwatchers, Peterson's passion ignited early. Born in Jamestown, New York, to European immigrants (Swedish father, German mother), Peterson was fond of saying that he couldn't remember a time when he didn't watch birds. He was small ("the runt of the primary school clutch" he wrote), shy and bookish, and found comfort and a calling in the natural world and particularly its birdlife. Getting little encouragement from his practical-minded family and open derision from his peers, Peterson found his first supporter and mentor in a seventh-grade teacher, Miss Blanche Hornbeck, whom he unfailingly credits in accounts of his amazing career for viewing his obsession sympathetically. The "pretty, red-headed" Miss Hornbeck organized a Junior Audubon Club and, armed with some leaflets from the National Audubon Society and a copy of Chester Reed's pocket *Bird Guides*, created a cell of young listers. As if being a birdwatcher wasn't bad enough, young Roger was also an artist (perhaps inheriting a creative instinct from his cabinetmaker father), and after an undistinguished

academic career in high school—where he was always "in trouble up to my scrawny neck" for drawing birds in his notebooks and contradicting his teachers on their bird facts—he lit out for New York City and a career in commercial art. While studying at the Art Students' League and the National Academy of Design, he began attending meetings of the Linnaean Society at the American Museum of Natural History. There—along with many others, but especially seven keen young members of the Bronx County Bird Club—he fell under the personal spell and peerless tutelage of Ludlow GRISCOM, the father of modern birdwatching.

In defining the nature of his immense achievement, Peterson was always at pains to point out that he did not invent the concept of "field marks" (Edward Howe Forbush may have introduced that idea), nor did he invent the reductionist approach to bird identification that focused on points of distinction (that was Griscom's genius). Rather what Peterson did—certainly better than anyone before him, and arguably at least as well as his many imitators—was to create a system of highly accessible painted images that showed at a glance how one species differed from another. The birds were intentionally depicted in a simple, almost diagrammatic, style and adorned with little arrows that pointed to the most useful field marks. The first edition of *A Field Guide to the Birds* was published by a risk-taking editor at Houghton Mifflin (Paul Brooks) in 1934 in the depths of the Great Depression—and, contrary to the expectations of many, soon became a best seller.

The success of the first field guide was, of course, just the beginning of a brilliant career that brought Peterson worldwide recognition not just as the bird man but also as the preeminent interpreter of the natural world. Through a field guide series that now numbers more than 60 titles and also as a lecturer, artist, photographer, and advocate for the natural world, Peterson not only was instrumental in introducing millions of people to the fascination of natural history but also helped to build the powerful constituency on which the conservation of the natural world depends.

In addition to his published works, Peterson's legacy is carried on at a thriving institute of natural history education that bears his name in Jamestown, New York.

Piracy

In an avian context, refers to the "stealing" of food from one bird by another—more technically known as "klepto-parasitism." The practice has been recorded in many species, including a variety of songbirds. Some raptors and seabirds obtain a significant proportion of their diet by harassing other birds to give up the prey they have caught, though all bird pirates are fully capable of catching or finding their own food. Where Bald Eagles and Ospreys occur together, the former will often pursue the latter until it drops the fish it has caught. Frigatebirds specialize in chasing boobies and other tropical seabirds and will even strike a bird that is reluctant to give up its catch. The food item does not have to be visible; pelagic pirates are aware that a pursued bird will disgorge swallowed food to "lighten its load" and make a faster escape. The most skillful maritime pirates habitually catch regurgitated tidbits before they hit the water.

In their distress—or perhaps in an attempt to soil their attacker or to relieve themselves of excess weight—the "victims" will sometimes defecate, and this has led to the belief among some seafarers that the pirates consume the excrement. This conviction has led to the jaegers and skuas being dubbed "jiddy hawks" (a version of an earthier term) and the notion has even been enshrined in the genus name of the jaegers and skuas *Stercorarius,* which means "eaters of excrement." While the chasers will not infrequently follow already-digested tidbits down by mistake, close observation confirms that the poop is rarely if ever ingested.

Arguably the most aggressive of bird pirates are the largest skuas formerly classified in a separate genus

(*Catharacta*) but now merged with smaller jaegers (*Stercorarius*). These powerful marine predators have been seen to grab gannets (birds twice the skuas' weight) by the wing, pull them into the sea, and try to push them under the water to force regurgitation.

Being among the most resourceful feeders, gulls have also mastered piratic techniques. In addition to chasing other birds, they have learned to accompany diving seabirds in the (often successful) hope of snatching a free meal; Laughing and Heermann's Gulls frequently stand on a Brown Pelican's head as an ideal perch for thievery.

Brown Pelican and
Laughing Gull

Plumage

The collective term for all the feathers that cover a bird's body. All birds eventually wear such a "feather coat," a characteristic that distinguishes them from all other classes of animal life.

COMPOSITION. The individual feathers, which together constitute the plumage, vary greatly in both shape and structure. Only major structural variations are described below.

Almost all the feathers normally visible on a bird, including the relatively large, stiff wing and tail feathers and the smaller, softer feathers that give a bird its smooth outline, are called the "contour feathers"; all of these have a central "vane" or "rachis."

Beneath the contour feathers there is usually a layer of short, soft, vaneless feathers present in varying degrees and in different locations among different species. These are either *down* feathers or *semiplumes*, which intergrade with each other and with types of contour feathers. However, down feathers are absent altogether in some species.

In addition to the contour feathers and the undercoat of down and semiplumes, there are *filoplumes*, long, narrow feathers, usually with a few "barbs" at the tip. These always grow with a contour feather and may be distributed over most of a bird's body but are difficult to see, unless you know what to look for and have really good optics (or the bird in hand).

Hardly recognizable as feathers are the *vibrissae*, stiff, hairlike feathers that appear as "eyelashes" in a few species of birds (e.g., Northern Harrier) but more often as *rictal* or *nasal bristles* around the top of the gape and over the nostrils, as in most flycatchers and night-jars.

Finally, there are the strongly modified feathers known as *powder down*. These are evenly distributed under the contour feathers of most birds, but in a few species, they occur in concentrated patches; they give off a fine dust made up of minute scalelike particles of keratin, the purpose of which is imperfectly understood.

To sum up: Contour feathers + down + semiplumes + filoplumes + vibrissae + powder down = plumage.

ARRANGEMENT OF FEATHERS. On casual inspection, a bird's body feathers seem to be evenly distributed, much in the way hair covers the human head. The reality, however, more resembles the head of the balding man who combs hair rooted in one place over a bare patch, for the body feathers of all but a few species of birds (e.g., ostriches, penguins) are rooted in discrete regions of the skin called feather tracts (*pterylae*), between which are open patches (*apteria*). The pattern made by the tracts varies consistently among different groups of birds and governs how the feathers are distributed. Typically, major tracts cover most of the head, throat, and neck; run down the center of the back in varying widths from neck to tail; surround the breast and belly; and cover the leading margins of the wings above and below and parts of the upper leg. The location and extent of the tracts obviously determine the location, extent, and number of the un-feathered areas. Down feathers may occur more or less evenly over the body, or be restricted to the apteria, or be absent altogether. Put another way, the apteria may be bare or covered thinly or thickly with down and semiplumes.

Normally, of course, the contour feathers lie over the apteria, so that, unlike most bald human pates, a bird's bare patches are effectively concealed.

HOW MANY FEATHERS? There are relatively few data on feather numbers (due in part, no doubt, to the tedious nature of the research), but a few basic tendencies have been established. As one would expect, small birds have fewer feathers than very large birds. The fewest contour feathers yet recorded is 940 for the Ruby-throated Hummingbird; the highest, 25,216 for the Tundra Swan. The numbers for passerines studied to date range between about 1,000 and almost 5,000, but, excluding a few extremes, the normal range seems to be between 1,500 and 3,000. Feathers are not distributed in equal numbers over a bird's body. The abundantly feathered swan cited above had 80% of its feathers concentrated on its head and very long neck. Small birds tend to have more feathers per square inch and in proportion to their body weight than large birds. This is consistent with the fact that smaller birds lose heat more rapidly and require more insulation. At least some species that must endure cold winters have up to 112% more feathers during that season.

FUNCTION. If we ask what benefits a bird derives from its feathers, the first answer likely to occur to us is: the ability to fly. Aerodynamically, plumage is the ideal body covering: it is light; it can be streamlined by compressing the body feathers to minimize drag; and the wing and tail feathers are flexible enough to allow high maneuverability, yet stiff enough to use wind power for lift and motion.

Control of body temperature, especially heat, may be an even more important function of plumage. The semiplume and down "underwear" provides insulation next to the skin, and the overlapping contour feathers can be "fluffed" to trap a maximum of warm air and "closed" to insulate swimming birds from cold water. Embryonic temperature and that of naked nestlings are also maintained by the natural insulation of parental plumage in addition to the "direct heat" applied through the brood patch. "Slippery" semiplumes and down may also serve as an "anti-friction" mechanism around the bases of wings and legs.

Plumage colors and patterns and adornments play essential roles in recognition, display, and protection against predators (see DISPLAY).

Female ducks, geese, and swans line their nests with down feathers plucked from their breasts, and some land birds (e.g., swallows) use "found" feathers both for lining and for decoration.

Feathers also serve as a binding material in the pellets regurgitated by bird-eating raptors, and grebes eat their own feathers, probably to protect the intestinal tract from sharp fish bones and spines.

Poetry

It is in poetry that birds and literature seem to form a natural alliance. The poet's need to conjure place and express emotion concisely but vividly finds a ready answer in birds' deep and varied symbolic (as well as real) presence in human experience. There is also a nice kinship between verse and birdsong that blends effectively with the poet's anthropomorphic metaphor in a poem such as Hardy's "The Darkling Thrush." Song may also be used to good effect in devising a poem's structure, as in the onomatopoeic lines in Whitman's "Out of the Cradle Endlessly Rocking," cited below. Finally, poets, as the ultimate wordsmiths, must appreciate the ample selection of useful and telling "sounds" present in the names of any avifauna. Words such as loon, albatross, eagle, plover, curlew, thrush, and sparrow serve the poet at least as well as they serve the taxonomist or birdwatcher.

Birds have been part of the tradition of Western poetry from its beginning. They appear in the lyrics of Homer (before 800 BC) and Catullus (first century BC); are mentioned in Beowulf, the first English epic (early eighth century); and by the end of the thirteenth century had fully assumed their archetypal function as symbols of renewal and the joy of nature in the famous anonymous rondel that begins

> *Sumer is icumen in,*
> *Lhude sing cuccu!*

From this point on, the English avifauna—particularly the Skylark (*Alauda arvensis*), the (Barn) Swallow (*Hirundo rustica*), the Eurasian Robin (*Erithacus rubecula*), the Common Nightingale (*Luscinia megarhynchos*), the Eurasian Blackbird (*Turdus merula*), and the Song Thrush (*Turdus philomelos*), as well as the Common Cuckoo (*Cuculus canorus*)—is generously represented in every period, form, and poetic usage, and in every length and quality of verse in the English language. From Chaucer to contemporary poets in England and America, the challenge is to find one who has not included a bird at least once.

Until well into the eighteenth century, the ornithological versatility of Shakespeare remained unchallenged—in

fact, it may remain so today, for evidence of which, see
SHAKESPEARE'S BIRDS.

But without doubt, the greatest literary bird refuge
of all was the age of Romantic poetry, beginning with
Blake's outrage at a caged Robin Redbreast (in "Augu-
ries of Innocence") and ending (perhaps) with Thomas
Hardy's pathetic darkling thrush. The Romantics are
characterized by their love of nature's rich, wild imagery
and a strong sense of "song," so it is hardly surprising to
find birdlife abundant in their poetry or that the period
produced the best-known (and perhaps best) "bird
poems" of all time: Shelley's "To a Skylark" and Keats's
"Ode to a Nightingale." Also hatched between the late
eighteenth century and the death of Queen Victoria
were Coleridge's deliverance-bearing albatross (in "The
Rime of the Ancient Mariner"); Burns's "green-crested
lapwing" ("Afton Water"); a well-stocked and songful
(but largely unlabeled) variety from Wordsworth; Tenny-
son's thrush ("Throstle"); Browning's unlikely gannet;
Swinburne's "sister swallow" ("Itylus"); Gerard Manley
Hopkins' kestrel ("Windhover"), to name a very few.

In modern poetry, man seems to take precedence
over nature, but this is due more to a change of context
than to any marked shift in theme. After all, Shelley's
skylark and Keats's nightingale are never glimpsed.
They are simply metaphors for an unblemished—totally
anthropomorphic—ecstasy, which is contrasted with the
melancholy that inevitably tempers man's joy. The tendency among many modern poets is to omit nature and
look at man in contexts of his own making—contexts
such as war and cities, which are generally inhospitable
to birdlife.

Nature is hard for a poet to ignore, however, and so we have Yeats's exquisite "The Wild Swans at Coole," the very rich and frequent avian imagery in Dylan Thomas (e.g., "Over Sir John's Hill"), T. S. Eliot's Sweeney placed ironically among Keats's Nightingales, and many other examples.

In examining American "bird poetry," it is hard to disagree with Welker (1955), who found little to praise. Most is sentimental doggerel—of the kind once published in newspapers: *Our merry friend the chickadee/ Chirps his song from the tall pine tree / Chick-a-dee-dee-dee, Chick-a-dee-dee-dee*, etc.

Poe's "The Raven" must be noted in passing, though the bird of the title is a pet and a literary device, not the subject of the poem.

Admirers of Emerson and Longfellow are bound to bring up the former's "The Titmouse" and various avian musings of the latter, for example, "Birds of Passage": "I hear the beat / Of their pinions fleet."

But there is really only one nineteenth-century bird poem worth the name, Walt Whitman's "Out of the Cradle Endlessly Rocking." Whitman was both a serious student of natural history and (much more importantly in this context) a serious poet. And the mockingbird in "Out of the Cradle" is not "nature observed" or a banal metaphor for some transcendental human quality; it is a voice of knowledge calling a boy out of his innocence:

> O solitary me, listening—nevermore shall I cease
> perpetuating you . . .

And Whitman uses the rhythmic repetitions of the mocker's song in his lyric. Leaves of Grass contains many

Emily Dickinson and
Baltimore Oriole

birds, used skillfully to evoke the wilderness, but there is
only one "Out of the Cradle Endlessly Rocking." Other
first-rate American poets have used avian themes or images, but few are cited among their makers' best work.
The poetry-reading naturalist may want to look up the
following, if she does not know them; these are not, with
a few exceptions, poems about birds, but serious poetry
that use bird imagery in a variety of ways.

— Emily Dickinson: "At Half Past Three a Single Bird";
   "The Robin Is the One"; "To Hear an Oriole Sing"; "I
   Dreaded That First Robin So"; "A Bird Came Down
   the Walk"; and many other poems with bird themes.
— T. S. Eliot: "Cape Ann" (the last part of *Landscapes*).

— Robert Frost: "The Oven Bird"; "Looking for a Sunset Bird in Winter"; "A Minor Bird"; "On a Bird Singing in Its Sleep."

— Vladimir Nabokov: "Pale Fire" (first stanza) in *Pale Fire*; not, of course, a "bird poem."

— Mary Oliver's poetry is suffused with images of nature including birds. See for example her collection *Red Bird*. Far from being sentimental, her work often has a dark or ironic cast, as in the following poem: "Showing the Birds"—

> Look Children, here is the shy,
> Flightless dodo, the many-colored
> Pigeon named the passenger, the
> Great auk, the Eskimo curlew, the
> Woodpecker called the Lord God Bird,
> The . . .
> Come, children, hurry—there are so many
> More wonderful things to show you in
> The museum's dark drawers.

— Wallace Stevens: "Thirteen Ways of Looking at a Blackbird" and "The Bird with the Coppery, Keen Claws."

— Robert Penn Warren: "Audubon" and "Ornithology in a World of Flux" in *Some Quiet, Plain Poems*.

— Ted Hughes: "Curlews Lift"; "Hawk in the Rain"; "King of Carrion" (in the *Crow* series, following the death of wife Sylvia Plath); "Hawk Roosting." See also FICTION; SHAKESPEARE'S BIRDS.

Politics, Birds in

There are at least three famous instances. During the investigation by the House Committee on Un-American

Activities of former State Department official Alger Hiss in 1948 as an alleged Communist subversive, it was crucial to establish the veracity of the testimony of Whitaker Chambers, himself a former Communist of questionable character. Chambers claimed intimate association with Hiss, and among many other corroborating details, noted that Hiss was an enthusiastic birdwatcher and had once been excited at having seen a Prothonotary Warbler along the Potomac River. By feigning casual good fellowship, in subsequent interrogation, Richard Nixon and U.S. Representative John McDowell (R-Pa), the latter also a birdwatcher, elicited from Hiss a confirmation of his encounter with the species and excitement over it. "Beautiful yellow head," Hiss exclaimed, "a gorgeous bird." He was ultimately convicted on two counts of perjury and sentenced to 10 years in jail.

Another instance involved the arrest of the antiwar activist and Jesuit priest Daniel Berrigan who, along with his brother Phillip, was among the leaders of the protest movement against the Vietnam War. Sentenced to three years in prison, he fled but was tracked to Block Island, Rhode Island—known as a birding hot spot during fall migration—by FBI agents, who posed as birdwatchers in order to allay suspicion. But Berrigan saw through their ploy and turned himself in—explaining to the officers that it was not yet birding season on the island. He served 18 months in prison and then continued his protests.

In March 2016 a female House Finch flew onto the stage and then hopped onto Bernie Sanders' podium during a rally at a packed stadium in Portland, Oregon, and sat for a few moments staring at Bernie as the crowd roared; the bird was dubbed "Birdie Sanders." (There are videos of the encounter on YouTube).

Poop, etc.

In birds, urine and feces are voided together through a single anal opening, the cloaca or vent, which is also use for copulation and egg laying. As anyone who parks a car under a starling roost or near a seabird colony can attest, bird droppings usually consist of solid dark wastes (feces) in a white, chalky semisolid material. The latter is urine, in the form of uric acid from which the water has been removed in the birds' kidneys and ureters. A few species, for example, geese and grouse, produce more solid, fibrous turds, the distinctive form of which is a reliable indication of the birds' presence.

Nests occupied by young birds over a period of weeks have an obvious potential sanitation problem. Birds use three basic approaches to addressing it. (1) A few species, for example, kingfishers and some woodpeckers (among other cavity nesters) are content to let their offspring wallow in their own wastes; despite the unwholesomeness we naturally attach to this "solution," it does not seem to affect the reproductive success of the species that practice it (however, see #3). (2) Most non-passerine nestlings quickly develop the habit of defecating over the edge of the nest or from the entrance of the nest hole. (3) An avian anatomical feature that human parents might envy is the *fecal sac*, a gelatinous pouch into which the feces of nestling (mainly passerine) birds are excreted. The adaptation would seem to be an evolutionary response to the survival advantages of nest sanitation. The parent birds remove the sacs from the nestlings' cloacae and either carry them away from the nest or (and here the human analogy breaks down) eat them. Some species habitually drop fecal sacs into bodies of water—including swimming pools.

Finally, the answer to the question everyone is asking: *Do birds fart?* According to Mark Murray, veterinarian at the Monterey Aquarium in California, it's not that they can't, it's just that they don't need to. That is, birds don't carry the kinds of gas-forming bacteria in their gut that humans and other mammals do to help them digest food, so there's no gas buildup to let loose. If gas did build up in their gut, there's nothing in birds' internal plumbing to prevent venting it. Murray notes helpfully that parrots sometimes emit fartlike sounds ("raspberries") to get attention, but that these are coming from the beak end, not the tail end.

See also GUANO.

Preening

The cleaning, manipulation, and arrangement of individual feathers of a bird's plumage using the bill; in the broadest sense, it includes the application of oil from the "preen (or uropygial) gland," a bilobed structure at the base of the tail present in most birds, from which birds can access an oily secretion with their bills.

In the course of normal activities, a bird's plumage becomes dirty, wet, matted with old preen oil, and infested with ECTOPARASITES, and the barbs of individual feathers become separated. All are potentially hazardous to the health and well-being of a bird (see PLUMAGE, Function), and preening is the innate activity—performed regularly and frequently—by which these conditions are rectified.

The basic preening action involves grasping an individual feather at its base and "nibbling" along its length to the tip or simply drawing the bill along the feather in a single, less meticulous action. This reattaches separated barbs, removes water and dirt, and in some instances applies oil. During preening, birds often come upon ectoparasites, which they usually seize and eat. The process involves all parts of the plumage and requires birds to assume an amusing variety of contorted postures. Swimming birds that preen while sitting on the surface execute what is known as a "rolling preen," in which they turn over on their side while floating with one leg in the air and preen their belly with their bill, apparently in perfect comfort. The one area of the body which birds cannot reach with their own bill is the head. This is preened by scratching with the feet (nails), by rubbing with the wing, or with the assistance of another bird ("allopreening").

Birds often engage in prolonged preening sessions during which—barring disturbance—the whole plumage is tended to, virtually feather by feather. A complete preening operation of this kind often follows other plumage-maintenance activities, such as bathing, DUSTING, SUNNING, or ANTING.

Prothonotary

The chief clerk or notary of the English court system and in some U.S. state courts. Also one of seven members of the College of Prothonotaries Apostolic of the

Roman Catholic Church, who concern themselves with recording canonizations, signing papal bulls, and other clerical responsibilities. The official robe of an English prothonotary includes a saffron yellow cowl and therefore provides an apt metaphor for the brilliant golden head and breast plumage of the Prothonotary Warbler (*Prothonotaria citrea*). Pronunciation depends on the usage: the clerk's and clerical titles are pronounced PROH-thuh-NO-tuh-ree, while the wood warbler's name is pronounced pruh-THON-uh-tary.

Upon hearing the title of the court official, U.S. President Harry Truman reputedly exclaimed: "What the hell is a prothonotary?" but later pronounced it the most impressive-sounding political title in the U.S.

Radar

The development of increasingly powerful and sophisticated radar technology beginning during the Second World War (1940) has greatly advanced our knowledge of specific details of nocturnal migration, which were once undetectable.

Essentially, radar (RAdio Detection and Ranging) involves sending into the air very high-frequency radio waves that reflect off the surface of any object they encounter. The returning images are collected by large (e.g., 28-inch diameter) parabolic dish antennas and projected onto a viewing screen.

Radar was developed, of course, to detect enemy aircraft at long range and at night, but even in the earliest days, an unexpected type of "echo" was picked up. These "angels," as radar technicians named them, were eventually proven to be migrating birds.

Using this system to detect high-flying nocturnal migrants (see MIGRATION), ornithologists have been able to observe and count for the first time the number of birds involved in migratory movements (far greater than previously suspected); to detect specific directions and routes taken (often more varied than imagined); and to obtain accurate data on timing, speed, elevation, and correlation with weather patterns. It is even possible to identify species using characteristic "echo signatures." When combined with ground observation of migrants in the daytime, radar information allows a close monitoring of migratory activity.

The history of migration study using radar—for example, Sidney Gauthreaux's long-term pioneering research on trans-Gulf of Mexico migration—is an exciting one to read because of the scope of the phenomenon and the continual succession of unexpected revelations. And anyone who has an opportunity to watch a radar screen fill up with the illuminated "shadows" of small birds on a night of heavy migration can experience one of the great benign miracles of modern technology; it is now possible to do this online!

Ratites

Flightless birds that lack a keel down the center of the breastbone, including among living species the ostriches, rheas, Emu, cassowaries, and kiwis. These were once classified as a single family, the Ratidae. With recent advances in biochemical research, the larger species are now classified in four different orders within the superorder Paleognathidae, whereas the smaller and very different kiwis turn out to be most closely related

to the extinct Giant Elephant-bird (*Aepyornis maximus*) and related extinct elephant-bird species of Madagascar. Modern species of ratites are native to Africa (ostriches); Australia and New Guinea (Emu and cassowaries); and South America (rheas); the five kiwi species are all endemic to New Zealand. Thus, ratites are absent from Asia and North and Middle America. However, when one "zooms out" phylogenetically, to the level of "clade," the clade Paleognathae includes not just ratites but also the tinamous—terrestrial birds of the New World that are capable of flight but typically walk rather than fly— and also the extinct, gigantic moas of New Zealand. Paleognathae means "ancient jaw." All other currently living birds—more than 10,000 species—belong to clade Neognathae, or "new jaw."

Religion

In Islam, Judaism, and Christianity, man is seen as a superior being, made in the image of the one true God. Birds and other animals are therefore traditionally held in rather low esteem, at best thought to have been created for human benefit, and at worst deemed "unclean":

some 20 proscribed species are noted in Leviticus and Deuteronomy. In this light, it is perhaps not surprising that in Islamic, Judaic, and Christian countries, birds have been recklessly exploited for food and sport, though conservation efforts have stemmed some of the excesses in the most prosperous and educated societies.

By contrast, the Hindu and Buddhist traditions espouse reverence for all life and consider human life simply a part (and a very transitory one) of a living whole. As a result of this less anthropocentric tradition, wild birds in India sometimes exist unmolested alongside burgeoning cities and villages, many of whose people are in need of food but will not sacrifice the life of a fellow creature to get it. Certain birds, for example, cranes, are the objects of particular reverence throughout most of Asia.

The Hindu-Buddhist traditions hold that we pass through a series of lives, in each of which we learn something of value to guide us ultimately to the blissful nonexistence called Nirvana in Sanskrit. For example, a soul might pass through one life as a falconer, the next as a falcon, and the next as a teal, each experience add-

ing a measure of enlightenment. Alas, birds are still in serious trouble in much of Asia due to the needs of the continent's enormous (and expanding) human population for the same resources the birds require to survive.

Of the more than 500 bird species known in modern Israel, fewer than 40 are identifiable in the Bible. However, the Old Testament contains numerous passages in which birds play prominent roles for dramatic effect. They are the first animals mentioned in the description of the creation (Genesis, Chapter 1) and "fowl of the air" fly commonly throughout the first book of Moses. Noah sent a raven, a kingfisher, and a dove to look for the land that would signal the subsiding of the waters and of God's wrath. The last returned with an olive branch from the top of Mount Ararat. An English tradition says that the kingfisher (presumably Common Kingfisher) flew high into the sky, thus acquiring its blue back, but too near the sun, thus "scorching" its breast; and for its foolishness, Noah made it stay out on the roof of the Ark and catch its food from the water.

Lastly, it is churlish no doubt to criticize St. Matthew for his lack of ornithological expertise. But when he says in his New Testament Gospel (6:26) "Behold the fowls of the air: for they sow not, neither do they reap, *nor gather into barns* . . ." he has obviously overlooked *Hirundo rustica*.

Other bird species, while not specifically mentioned in the Bible, have come to be symbolic of certain biblical themes, and often appear as such in religious paintings:
— The *Eurasian Goldfinch*, because of its association with thistle (= thorns), came to stand for the passion of Christ, its face stained red with his blood;

— *Doves* represent purity as well as peace and are therefore often depicted in the company of the Virgin; the dove is also representative of the Holy Ghost or Spirit and thus was sent by the Father to inseminate the Virgin with the Son;

— The *eagle* is a symbol of the resurrection as well as the personal sign of Saint John the Evangelist;

— The *pelican* is sometimes associated with Christ because of the erroneous notion that it can feed its young with blood from its own breast;

— A number of bird species, for example, the *Barn Swallow*, have traditionally been credited with trying to ease Christ's suffering by pulling thorns from the crown or nails from the cross, drawing blood from their own (now reddish) breasts in the process; and ...

— The peculiar mandibles of the *crossbills* have been attributed to the pulling of crucifixion nails.

Roc (originally, Rukh)

A gigantic bird of Arabic legend, said to be able to carry off elephants in its talons. In the Western world, it is best known from the second and fifth voyages of Sinbad in *A Thousand and One Nights*. In the phylogeny of myth, the Roc is closely associated with other avian giants, such as the Anka (Arabia), Simurgh (Persia), and Phoenix (classical legend). It was thought to nest on an island in the Indian Ocean and eventually became identified with the Giant Elephant-bird of Madagascar (*Aepyornis maximus*), which, though flightless, was the heaviest bird known to exist (see SIZE and GIANT BIRDS) and laid eggs with a capacity of 2 gallons. The enormous fronds of the *Raphia* (*Sagus*) palm of Madagascar

were passed off (on the Great Khan among others) as feathers of the Roc.

RSPB (Royal Society for the Protection of Birds)
In the United Kingdom in the late nineteenth century, milliners and other commercial interests promoted the slaughter of millions of wild birds in order to use their plumage as adornments for hats and other articles of fashionable female attire. As an example of the scope of the slaughter, the skins of some 760,000 "exotic" birds were imported to Britain in just the first few months of 1884.

As a counterpoint to the many women who chose to wear expensive feather fashions, several women's organizations were founded to put an end to the carnage.

The most prominent of these were The Plumage League founded in 1889 by Emily Williamson, to campaign against the use of the skins of grebes and kittiwakes in the making of fur clothing, and The Fur, Fin, and Feather Folk, also founded in 1889 by Eliza Phillips, Etta Lemon, Catherine Hall, Hannah Poland, and others. These groups rapidly attracted large followings and in 1891 joined forces to form the Society for the Protection of Birds in London; the organization was granted its Royal Charter in 1901.

At first the members of the organization were all women, and women—notably many of high social status—have continued to be prominent in the RSPB's leadership. The Duchess of Portland, for example, was its first and longest-serving president. The Society's earliest mandate was:

— That Members shall discourage the wanton destruction of Birds and interest themselves generally in their protection; and

— That Lady-Members shall refrain from wearing the feathers of any bird not killed for purposes of food, the ostrich only excepted.

In 1906 the Society successfully petitioned the British Parliament to pass laws banning the use of plumage in clothing.

Today the RSPB is the largest environmental organization in Europe with more than a million members of all demographics, including 195,000 youth members. It manages 200 nature reserves in the UK and focuses on Action for Birds, including programs in research, education, and advocacy. With partners such as BirdLife International, its reach is global.

For a striking U.S. parallel in the development of a bird conservation organization, see HEMENWAY.

Seawatching

A form of relatively stationary birdwatching typically at a coastal point where large numbers of waterbirds of more than routine interest can be expected to pass at certain seasons and under particular weather conditions. (Lakewatching is similar, though typically requires more patience.) One difference between seawatching and that other birding subcategory, HAWKWATCHING, is that the latter is best done on a fine spring or fall day in some majestic scenic location such as a mountaintop or an escarpment above a river valley. On a promising seawatching day, by contrast, the weather is likely to be spectacularly foul—gale force winds howling and sheets of sleet streaming directly into your face and telescope. It is true that the sea can take on a dramatic appearance during such events, but it is often obscured by some form of precipitation. The bad weather has the effect of pushing bird species that normally shun the sight of land (e.g., shearwaters, jaegers, phalaropes) into viewing range, and if onshore winds coincide with migration, spectacular movements of loons, sea ducks, and alcids, often laced with rarities, sometimes pass well within binocular range. In another common seawatching scenario, the weather is quite good but the birds are scarce to absent; this requires a different kind of endurance.

The propensity to seawatch appears to be a reliable genetic marker that neatly divides the birding community. On one side are those (present author included) for whom there is no greater pleasure (a couple of exceptions) than to repose near the coast in the teeth of a cyclone watching masses of blurry avian forms passing through foggy oculars and keeping a sharp lookout for the Long-tailed Jaeger or Bridled Tern that is bound to show up if you stick it out a little longer. The other birder subspecies is exemplified by Bill Oddie, who calls seawatching "the most tedious of all pursuits" (including cricket!) and devotes an entire chapter of a book to excoriating the activity and its practitioners: "The confident expert will pick out dots beyond the normal range of human vision and tick them off as rare petrels or shearwaters."

Even if you were born seawatch-negative, you should endure the experience at least once, if for no other reason than to gain merit for the bestowal of future rarities.

There is also, of course, scientific seawatching for the purpose of gathering data on migratory species.

Sex

From the human point of view, the avian sex act is both bizarre and, well, anticlimactic. The male stands on the female's back; she lifts her tail; they both force their single anal orifice (see CLOACA) inside out, and the semen is transferred in what has been described as a "cloacal kiss." As may be imagined, it is difficult to observe the precise internal details of the procedure, but it may be that part of the male's cloaca is actually inserted into the female's oviduct for ejaculation. A few groups of birds (ostriches, kiwis, most waterfowl, some gallinaceous birds, and at least one songbird) have evolved a penis of sorts. The ostriches' "intromittent organ" is red and eight inches

i.

ii.

iii.

Mute Swans mating

long; and the phallus of the nine species of "stiff-tail" ducks (e.g., Ruddy and White-headed Ducks) is said to be heavily "ornamented." In most birds, however, the male member can be described as a sac of erectile tissue that is thrust out or retracted in a manner that has been compared to reversing the finger of a glove. Females of some species also possess a rudimentary clitoris. An explanation for the additional equipment in waterfowl is that it facilitates the transfer of sperm in the water.

Bird semen is similar to human semen in color and consistency. But while the quantity transferred in a single ejaculation is of course far less in birds than in humans, the concentration of spermatozoa is typically many times greater—up to 3 billion per ejaculation compared with the human average of 300–500 million.

Shakespeare's Birds

At least one scholar has taken the trouble to confirm that the Bard of Avon included more than 600 references to birds throughout the plays and sonnets, including more than 60 identifiable species. But more impressive than Shakespeare's "list" is the resourcefulness with which he makes use of his considerable knowledge of both the habits of common species and the folklore and superstition that surrounded them. He works dozens of changes on the familiar spring/love/birdsong theme, but also has birds appear as omens, jokes, character analogues, in genre scenes, and in examples of nature's simple wisdom. Two examples must suffice.

Having spent the night with Juliet, Romeo rises to leave her rooms before he is discovered.

> JULIET. Wilt thou be gone? It is not yet near day:
> It was the nightingale and not the lark,
> That pierced the fearful hollow of thine ear;
> Nightly she sings on yon pomegranate tree:
> Believe me, love, it was the nightingale.

> ROMEO. It was the lark, the herald of the morn.
> No nightingale: look love what envious streaks
> Do lace the severing clouds in yonder east:
> Night's candles are burnt out, and jocund day

Stands tiptoe on the misty mountain tops.
I must be gone and live or stay and die.
(But Juliet convinces him to stay, then realizes the danger)

JULIET. It is, it is: hie hence, be gone, away!
It is the lark that sings so out of tune,
Straining harsh discords and unpleasing sharps.
Some say the lark makes sweet division:
This doth not so, for she divideth us:
Some say the lark and loathed toad change eyes;
O, now I would they had changed voices too!

Romeo and Juliet, ACT III, SCENE V.

Waiting for the fatal engagement with the army of Octavius Caesar and Marc Antony on the Plains of Philipi, Cassius speaks to his friend Messala about avian omens:

You know that I held Epicurus strong,
And his opinion: now I change my mind,
And partly credit things that do presage.
Coming from Sardis, on our former ensign
Two mighty eagles fell; and there they perch'd,
Gorging and feeding from our soldiers' hands;
Who to Philipi here consorted us:
This morning are they fled away and gone;
And in their steads do ravens, crows and kites
Fly o'er our heads, and downward look on us,
As we were sickly prey: their shadows seem
A canopy most fatal, under which
Our army lies, ready to give up the ghost.

Julius Caesar, ACT V, SCENE I (See ORNITHOMANCY)

See also POETRY.

Size (length, height, weight, and wingspan)

LARGEST BIRDS KNOWN TO HAVE LIVED. The Giant
Moa (*Diornis maximus*) of New Zealand, a wingless
relative of the ostrich that stood as high as 13 feet (4 m)
with neck extended and weighed more than 500 pounds
(226.8 kg), is the *tallest* known bird. However, the Giant
Elephant-bird of Madagascar (*Aepornis titan*), which
stood a mere 9–10 feet (2.7–3.0 m) tall, is the *heaviest* known bird, weighing in at more than 950 pounds
(430.9 kg).

LARGEST LIVING BIRD is the Common Ostrich
(*Struthio camelus*), some of which stand 8 to 9 feet (2.4–
2.7 m) tall and weigh at least 345 pounds (156.5 kg) (the
newly recognized Somali Ostrich is probably similar).
Their living relatives, the cassowaries, emu, and rheas,
are also far from petite.

LARGEST FLYING BIRD KNOWN TO HAVE LIVED. A
recently discovered fossil of a prehistoric condor relative,
the Giant Teratorn (*Argentavis magnificens*), lived in the
late Miocene 6 to 8 million years ago in South America. It is estimated to have weighed about 180 pounds
(81.6 kg), but with a wingspan of between 20 and 26
feet (6.1–7.9 m). A rival for the wingspan record is another recently discovered teratorn, *Pelagornis sandersi*,
a denizen of the Oligocene some 25 million years ago
with an estimated wingspan of 20–24 feet (6.1–7.3 m). It
was unearthed amid construction work at the Charleston
(South Carolina) International Airport. Teratorn means
"monster bird."

LARGEST LIVING FLYING BIRDS are species of bustards, pelicans, swans, and condors. The Great Bustard
of Eurasia reaches at least 44 pounds (20 kg), and Mute

and Trumpeter Swans equal or exceed this weight—
though the ability of an alleged 49.5-pound (22.5 kg)
Mute Swan to become airborne has been officially
doubted.

LIVING BIRDS WITH THE GREATEST WINGSPANS.
The Marabou Stork may hold the absolute record with a
maximum measured wingspan of 12 feet (3.6 m), though
9 feet (2.7 m) is more usual for this species. However,
Wandering and Royal Albatrosses have *average* wingspans of more than 10 feet (3 m) (11 feet 4 inches; 3.45
m confirmed for Wandering), and some individuals may
attain a 13-foot (4 m) span.

SMALLEST LIVING BIRD is the Bee Hummingbird,
endemic to Cuba and the Isle of Pines, males measur-
ing 2¼ inches (5.7 cm) from bill tip to tail end and
weighing about 1/10 of an ounce (3 g).

See also GIANT BIRDS.

Ostrich and
Bee Hummingbird

Smell

Fishermen and birders have known for ages that cod livers or the like spread on the sea at a favorable place and season can rapidly attract flocks of seabirds from beyond the horizon and have assumed that they must be attracted by the smell. On the other hand, John James Audubon once colorfully claimed that Turkey Vultures found his painting of an eviscerated sheep more attractive than a genuine "ripe" carcass that he had hidden nearby.

As the avian sense of smell became a matter of scientific investigation, ornithologists initially tended to correlate birds' ability to detect odor with the size of their smelling organ (olfactory bulb) relative to that of the forebrain. The not-illogical assumption was that birds with big olfactory bulbs—notably kiwis, grebes, tube-nosed seabirds, New World vultures, rails, and nightjars—were accomplished smellers, and species with negligible smelling apparatus, for example, songbirds, had little or no ability with this sense. In the 1980s, however, a range of experiments testing neurological responses to strong odors and the ability of various birds to find food or nest sites when the olfactory apparatus was deliberately impaired or, conversely, a variety of smelly clues were provided, have altered that assumption. We now know, for example, that:

— The avian olfactory apparatus is in the same league with that of reptiles and mammals and that some birds can outperform these other classes in certain smelling tests. Storm-petrels for example can detect the gaseous emanations of krill from as far away as 25 km.

Storm-petrels locating their burrows by smell

— The olfactory bulb correlation was not wrong, just insufficient, and the birds that were assumed on this basis to have "good noses" (see above) actually do. The converse is not true, however, and it is now assumed that small songbirds with apparently weak equipment use smell routinely as a detection device.
— Birds that have been proven to have a good sense of smell fall into some logical categories, for example, birds that must search for food that is not readily visible over a wide area (seabirds, vultures); birds that must find food or nest sites in the dark or in dense habitats (kiwis, tubenoses, rails, nightjars); and species with special requirements that are detectable by smell (honeyguides and beeswax).

— At least some male ducks are attracted to sexual odors given off by females in the breeding season.

— In some instances, smell is of prime importance in locating nest sites, for example, petrel burrows that must be located at night in dense forest. This homing function may also be more important that we thought in at least some land birds, for example, pigeons.

— Birds can discriminate among plant as well as animal smells and use this ability to choose appropriate food items and nesting materials.

While there remains a consensus that for most birds, vision and hearing are more important than smell in living their lives successfully, it is now apparent that for some groups of birds, the precise detection of odors is critical and that, overall, birds smell a lot better than we thought.

For how birds smell to humans, see ODOR.

Soaring

by birds can be defined as the ability to move through the air without flapping; there are two basic types.

THERMAL SOARING depends on columns of rising warm air (thermals) that are generated at irregular intervals from the Earth's service via solar heat convection. As the column rises, it expands toward the top and is finally set free by surrounding cold air. (Picture a bubble of air rising from the bottom of a glass container of boiling water.) Within the rising bubbles there is a revolving circle of warm air with a continual updraft of cold air through the center of the column. Locating one of these thermals, certain birds can "ride" the rising ring of moving air to the desired height and then drift off in the desired direction, finding another thermal when they lose altitude. This explains the hawk "making lazy circles in the sky." Birds that use thermals habitually typically have broad wings and short wide tails, enabling them to "catch" maximum lift. These include many raptors as well as pelicans, cranes, and storks, but narrow-winged species such as gulls and swifts also use the technique to cover long distances in search of food or during migration without expending too much energy. See also KETTLE.

DYNAMIC SOARING exploits the fact that winds blowing over the sea are slowed by the waves at the surface and gradually increase in velocity with altitude. Relatively heavy birds with long, narrow wings, such as albatrosses, can gain speed high in the "fastest" air and then plunge downward across the wind; when they reach the slower air near the sea surface, they use their momentum to head up again, simultaneously turning into the wind which blows them back aloft—all without a single flap.

Soft Parts

Exposed, unfeathered parts of a bird's body, including the bill, legs and feet, fleshy eye-rings, facial skin, combs, wattles, and irises. In some species, soft part colors can change rapidly, becoming more intense during courtship or the breeding season and fading to a subtler hue during the rest of the year. At any season, soft part colors fade after death and therefore must be noted on the labels of museum specimens. The name is not very apt, since, excepting the skeleton, bill, and claws, many of the "soft" parts are in fact among the hardest parts of a bird's body. Soft part colors are sometimes useful field marks for distinguishing between similar species.

Song

To the human sensibility, birdsong is mainly a source of aesthetic delight. It is a major component of our general concept of "nature." It may even have given rise to our realization—not reliable in all cases—that we too can sing.

From a bird's perspective, of course, the sounds it makes have an altogether different significance. They are a practical means of communication and expression, which for most species are as necessary for survival as visual signals such as color, pattern, and physical gesture (see DISPLAY).

It is not surprising then that bird sounds are by now highly evolved. Although sound plays an important role in the lives of many insects, amphibians, and mammals, only human speech and the vocalizations of some cetaceans surpass bird sounds for vocal subtlety and complexity.

WHAT IS A "SONGBIRD"? In scientific classification, the songbirds make up a suborder (Oscines) of the large cosmopolitan order of "perching birds" (Passeriformes). This taxonomic distinction is based on the relative complexity of vocal anatomy among the Oscines (see below), but it is not a reliable indicator of vocal virtuosity or finesse. While it is true that some of our finest songsters, for example, the thrushes, belong to the Oscines, so do the "plain spoken" crows and jays. Conversely, virtuosi among the shorebirds, owls, and nightjars are not in the same order as the "official" songbirds.

HOW DO BIRDS SING? Regardless of how the singers are classified, virtually all vocal bird sounds are produced in an organ called the *syrinx*, which is unique to birds. The human voice is produced in the larynx (or "voice box"), a modification of the upper part of the windpipe (trachea), which holds the vocal chords. In birds, the upper part of the trachea is also called the larynx, but it contains no vocal chords and does not produce sound. In most birds, the syrinx is located at the juncture of the trachea and the bronchi.

Attached to the syrinx are pairs of muscles that control the quality of sound production. The Oscines have up to nine pairs, whereas most other birds make do with just one or two pairs. Within the syrinx are elastic membranes that can be stretched and relaxed by air pressure and by the workings of the syringeal muscles. The membranes can also be manipulated within the air passages of the bronchi to regulate the amount of air passing through. When air is passed through the syrinx from the lungs, the membranes are made to vibrate, producing a sound in the way that a blade of grass held between the thumbs will "whine" if you blow across it. The sound can be modified (1) by stretching or loosening the membranes (for higher or lower pitch); (2) by making them vibrate slowly or rapidly (for tonal qualities); (3) by altering the direction of the passage of air (for loudness or softness); and (4) by starting and stopping air flow (for rhythm). Unlike human speech, birdsong is not inflected much if at all by resonating in nasal, mouth, or throat cavities. This is dramatized by the ability of many species to sing full, rich songs with their mouths filled or their bills closed.

Perhaps the most remarkable feature of a bird's voice is the ability to sing through both bronchi simultaneously or separately. This "twin voice" is responsible for the harmonic sounds present in most songs and can also produce *two distinct themes at the same time*, giving songs like those of some New World thrushes their ethereal quality.

WHY DO BIRDS SING? Not long ago it would have seemed pointless to ask *why* birds sing. But since Darwin, the inquisitive have been schooled to realize that the presence of such a conspicuous trait in so many successful species must imply significant "reproductive advantage," since elaborate song seems to be a major factor in mate choice, sexual selection, and competition among males. That there must be advantages in making loud noises is further implied by very apparent disadvantages. What better way, after all, to attract the attention of predators than to sing a lengthy song over and over from a prominent perch?

Territorial Song. Most of the birdsongs we hear are those of male birds advertising their presence on their territories to prospective mates and potential rivals of their own sex and species. Neighboring males engage in singing contests to establish their invisible boundary lines. And they may escalate their song battles by adding aggressive motifs, but they usually come to terms vocally without resorting to physical altercation apart from the occasional chase-off. Wandering males are also warned efficiently when they trespass on occupied real estate. This territorial song usually continues through the breeding cycle and seems to cement the pair bond after mating.

Emotional Release Song is usually very distinct in phrasing from territorial song and seems to represent a pure release of energy of the kind that Shelley attributed to his skylark. Songs of this kind have been recorded for many passerine species and may be accompanied by an "ecstasy flight" (not to be confused with territorial song flights characteristic of many open country species). In some species (e.g., some tyrannid flycatchers), they are likely to be performed at twilight, but they may also be given at night during migration or on the wintering ground. They often contain many improvised elements. The function of this category of song is unclear. But sober scientists, who eschew anthropomorphism by profession, acknowledge its expressive quality and have even suggested that it contains a germ of artistic invention!

Female Song. On the whole, the singing of a full (i.e., territorial-type) song is unusual among female birds, and it has even been demonstrated that there is some causal relationship between sex hormones and the development of the vocal mechanism (syrinx). Females that *do* sing tend to do so much less frequently than the males of their species and only in specific situations. It has become clear through field research, however, that female song is by no means rare and is practiced in some form by members of a wide variety of families. In some groups (e.g., the thrushes and cardinal-grosbeaks), the female's song can be equal to her mate's in complexity and intensity.

HOW OFTEN? The Red-eyed Vireo of North America with its short, whistled phrase, uttered seemingly without pause throughout the day, appears to hold the record for the most songs sung per unit of time:

22,197 in 10 hours; for its tireless efforts the species has been dubbed The Preacher by some human listeners. Other such patiently gathered statistics indicate that the norm for passerine species is closer to 1,000 to 2,500 songs per day.

VOLUME. In certain circumstances, birds are known to sing extraordinarily softly. This may simply be a *sotto voce* rendition of the territorial song (so-called whisper song) or may be completely distinct. The different variations are often lumped together as "subsongs" or "secondary songs" and are often lower in pitch and more extended as well as lower in volume than ordinary songs. They are sometimes sung during bad weather or during the hottest part of the day.

At the other end of the sound spectrum, researchers in Brazil reported in the October 2019 issue of *Current Biology* that the male White Bellbird (*Procnias albus*), native to the Amazon Basin, has broken the record for the *loudest bird song on record*, reaching 125 decibels on average at the loudest point in its songs, edging out a closely related South American species, the Screaming Piha (*Lipaugis albus*) whose "song" hits maximum volume at a mere 116 decibels. For context, 110–125 decibels is technically defined as "extremely loud," the level at which other sounds can't readily be heard, for example, an airplane taking off at close range. At 125, the sound is technically described as "painful." A logical truism about the volume of bird song is that birds that live in dense habitats have the loudest songs. But the bellbird, whose sound might be likened to a resonating gong, sings from an open perch and ideally right in the face of its pitiable potential mate.

White Bellbird pair

CALLS AND OTHERS SOUNDS. Calls are typically brief as befits their function of conveying immediate information within the routine activities of a bird's daily life. There is great variation among species in the richness of call vocabulary. All birds are capable of producing some kind of call, if only a hiss or grunt elicited under direst stress. Many species (e.g., grebes, herons) have no true song as defined above but produce a fair diversity of calls, especially on the breeding grounds—though the precise import of each of these sounds has not been defined for many species. Songbirds (see above) are the most articulate callers, with more than 20 distinct vocalizations recorded for several species.

Among the commonest functions of calls are: defense, warning, distress, begging (by nestlings and juvenile birds), flock cohesion (e.g., on migration), identification of a food source (e.g., in foraging birds) or a predator (see MOBBING), gathering (e.g., by parents of wandering broods), and comfort. Certain calls may be used exclusively by birds of a certain age, sex, or species, while others may be used and recognized among many different species.

Nonvocal sounds are also used by many species to convey a variety of messages. Bitterns, many grouse species, and a few sandpipers make unique naturally amplified hooting sounds by forcing air in and out of their inflatable esophagus, greatly enhancing the distance over which these eerie sounds can be heard. Bills become percussion instruments when snapped or struck against each other, as with albatrosses, and the meaningful sounds made by air passing through specially evolved groups of feathers, as in the woodcocks and snipes, are wonderfully diverse.

See also DUETTING; VOCAL MIMICRY.

Speed

In normal traveling flight, most birds progress at between 20 and 50 miles per hour (32.2–80.5 kph) but are generally capable of higher speeds, when, for example, they are being chased by a predator. Racing pigeons and Red-breasted Mergansers have been reliably clocked at air speeds of 80 mph (128.7 kph), and doubtless many other species have a similar capacity.

There are reports that the fastest birds may attain speeds reaching or exceeding 200 mph (321.9 kph); however, such speeds have never been recorded in a controlled scientific procedure. Many velocity records come from pilots who check their speedometers as their planes are passed by birds. Dunlins, for example, have been clocked from a plane at "not less than 110 mph (177 kph)."

Some alleged and confirmed speed records are as follows:

FASTEST BIRD WORLDWIDE (flapping flight). This record *may* belong to the White-throated Needletail (*Hirundapus caudacutus*), an Asian swift species whose speed has been reliably clocked at 106.25 mph (171 kph); the same species is also said to have reached 219.5 mph (353.3 kph) (ground speed), but the method of recording this statistic has been called into question. A similar record *might* belong to the White-throated Swift of North America, which has been "estimated" to attain 200 mph (321.9 kph). (There's a reason that they're called "swifts.")

White-throated Swift and Peregrine Falcon

One of the most persistent popular myths about speedy birds is that the world record belongs to the Peregrine Falcon, which "stoops" on its prey by folding its wings and dropping from a height. In this (nonflapping) "flight," Peregrines have long been credited with speeds of at least 175 mph (281.6 kph) and possibly more than 200 mph (321.9 kph) by several pilots; however, researchers who recently attached small air speedometers to this species were unable to confirm a diving speed greater than 82 mph (132 kph).

FASTEST RUNNING BIRDS. The Common Ostrich of Africa holds the world record, which is at least 44 mph (70.8 kph) and may reach 50 or 60 mph (80.5–96.6 kph). Ring-necked Pheasants are said to be capable of 21 mph (33.8 kph), and Roadrunners and Wild Turkeys 15 mph (24.1 kph).

FASTEST SWIMMING BIRD. The Gentoo Penguin has been timed underwater at 22.3 mph (35.9 kph).

Sunning

Many species of birds are known to assume "peculiar" postures, spreading and fluffing their feathers to expose their plumage to the light and/or heat of the sun. Though the behavior is sometimes observed on cool days, a number of witnesses record that the assumption of a sunning posture seems to be triggered when a bird suddenly experiences a rise in heat (and light?) intensity.

Sunning behavior varies among species and apparently also according to the sun's intensity. Among songbirds, a typical posture involves squatting on the ground at a right angle or facing away from the sun with wings drooped or outspread, tail fanned, and body feathers (especially those of the head and tail) erected, that is, fluffed. In this position the bird's head, neck, mantle, rump, and the upper surfaces of wings and tail receive the full impact of the sun's rays. Another posture, assumed by some swallows, pigeons, and others, is to roll over on one side, raise the wing, and expose the undersurface to the sun. Many waterbirds simply stand with their backs to the sun, perhaps with their wings drooped or necks stretched. Especially when the heat is most intense, sunning birds frequently open their bills and pant (see AIR CONDITIONING). As in water bathing and DUSTING, birds often have favorite sunning spots to which they return regularly.

While sunning, birds often seem to go into a kind of trance, allowing people to approach much more closely than normal. The bizarre posture and behavior strongly suggest to the human observer that sunning birds are injured—or perhaps suffering from sunstroke!

The main theories advanced to date to explain sun-bathing in birds are (1) that exposure to heat and light activates ECTOPARASITES such as bird lice and perhaps drives them from areas of the body the bird has most trouble reaching and/or to areas where they can be captured most easily in the bill; (2) that the sun's ultraviolet rays release vitamin D from the preen oil, which in turn is ingested by the bird in the preening that typically follows a sunbath; (3) that the sun dries and fluffs the feathers by evaporating moisture and oils from the plumage (as may be true of dusting), thus maintaining good insulation; (4) that birds may be able to increase energy reserves by absorbing solar radiation directly through the skin; (5) that it feels good, especially when molting causes skin irritations; and, of course, (6) that it is some optimal combination of 1–5.

For related behavior, see PREENING; DUSTING; ANTING.

Swan Song

Debate about the nature—or very existence—of a unique "song" uttered by some swan species with their last breath has raged at least since classical times. Plato did not doubt that swans sang in their last hour but argued that this was simply a final affirmation of their cheerful, musical disposition—a description that few present-day swan watchers would recognize. Other commentators have contended that the swan song is as mythical as the lustful cygnine that ravished Leda.

There can be no doubt that swans are capable of a variety of vocalizations, most notably "bugling" or "trumpeting" notes, possibly facilitated by their unusually long, coiled tracheas, but also muted "conversational" sounds and, especially in the Tundra Swan, pleasing, musical tones; pairs of this species are known to engage in prolonged "duets" in which the two sexes take different parts. Delacour (1954) opines that the swan song is simply (unpoetically) the expulsion of air from the trachea of a shot swan as it plummets to Earth, but there are several authoritative accounts of an extended, melodious, but rarely heard call made by Tundra Swans, sometimes (but not always) during their death agony. The waterfowl expert A. H. Hochbaum described what he calls a "departure song"—a soft, beautiful series of notes uttered before takeoff— and this description agrees well with the account of the nineteenth-century naturalist D. G. Elliot, who recorded a similar "song" from a Tundra Swan he brought down out of the North Carolina sky.

In vernacular usage, the swan song is a metaphor for a last earthly act, usually a creative one, for example, a poet's last sonnet or an actor's last role. Even a politician's last speech may be so described by the broad-minded.

Tameness

Like other "wild" animals, birds are generally wary of close contact with human beings, yet the degree to which some species or individual birds will tolerate our company is by no means uniform, and cases of extraordinary tameness are common in the literature. In many cases it is clear that birds sometimes relinquish much of their wariness not from any desire for intimate contact with people themselves, but rather because human society makes food and shelter available either inadvertently (cleaning fish) or by design (birdhouses). Yet there are numerous other instances of both fear and tameness in birds that cannot be explained as learned behavior and that therefore seem to have some genetic basis. So many phenomena have been cited under the heading of "tameness," and so little effort made so far to relate one to another, that it seems best simply to list some of the more conspicuous facts in the matter.

— Birds that are wary of humans *per se* often overcome their seemingly innate fear when compensation in the form of food or nesting sites is available. There are innumerable examples: gulls coming on board fishing boats; gulls, crows, and other birds following

on the heels of a plowman; Barn Swallows, phoebes, finches, robins, wrens, and others nesting over
a much-used doorway or inside someone's porch;
Gray Jays raiding campers' rations and even taking
food from the hand; and, of course, the enthusiastic
response of many species to being fed store-bought
seeds and other "bird food."

— Certain families and species of birds seem to be
"naturally" tamer than others and need no training
by or experience with humans in the matter. Many
seabirds, grouse, owls, shorebirds, corvids, tits, kinglets, waxwings, and "winter finches" seem not to recognize a human being as a threat and are thought
variously to be "friendly," "bold," or "stupid" by the
objects of their trust.

— Many of the species and individuals in the preceding groups breed in remote areas of the world
where their contact with humans can be presumed
to be limited. The birds (and other animals) of the
Galapagos and other remote oceanic islands are notably fearless. The species of grouse, owls, jays, waxwings, and finches that have the greatest reputation
for tameness breed largely in remote boreal forests,
while many shorebirds nest in a vast wilderness of
uninhabited tundra.

— There are numerous cases of apparently inexplicable
geographical variation in tameness within a single
species or among like ones. American birdwatchers
are always struck by the reclusiveness of many wood-
land birds (e.g., woodpeckers) in the British Isles
as compared with the United States—where perse-
cution of songbirds has if anything been far greater

than Britain's in recent times—or with other parts of Eurasia, where ancient traditions of trapping song-birds for food are still intact.

— There is much variation in wariness among individuals of a species: not all Pine Grosbeaks may be closely approached; American Robins breeding in forested areas are likely to be shier than their suburban counterparts.

— It has been demonstrated that organisms tend to have an innate negative response to sizes, shapes, and movements that correspond to those of their natural predators. Some species may be genetically "programmed" to respond in this way to humans, whereas others may have to learn this fear. In this connection it is interesting that birds that are very wary of a person on foot are often not nearly so shy when approached in a car or on horseback.

— To the extent that this can be fairly judged, some individual birds seem to "enjoy" human company over and above the practical benefits (food) they derive therefrom. There are numerous records of birds in captivity (e.g., parrots) being petted and ultimately seeking out their captor and "begging" to be stroked. This may be referable to some notably social species' urge to preen other birds and be preened.

— The "tameness" of birds that stay on their nests and allow themselves to be touched or that attack a human being who trespasses on breeding territory is probably best interpreted as defensive behavior, which represents a distinctly different response to the human presence from the other reactions discussed here.

—The "tameness" of birds that become imprinted by a human at hatching (see INTELLIGENCE) should probably also be considered as a separate phenomenon, due to its abnormality.

Taste (Sense of in birds)

In humans, taste is a sophisticated sensation blending five basic elements—salty, sour, bitter, sweet, and savory—in varying strengths combined with a much broader range of perceptions received through our sense of smell. The main organs of taste in all vertebrate animals are collections of cells—the taste buds—that send and receive signals to and from the brain via aggregations of nerve fibers. Human taste buds are located mainly on the tongue—where they are readily visible to the naked eye—and in the wall of the pharynx (upper throat).

The taste sense in birds is much simpler structurally, with an average of only 30–70 inconspicuous taste buds (humans have about 9,000) located mainly in the throat and the softer areas of the palate, with usually only a few in the relatively soft back of the tongue; taste buds have also been detected in the bill edges of some species. Among the exceptional aspects of the large fleshy tongues of parrots is the presence of some 400 taste buds.

The physiology of taste is not completely understood in humans and is little studied in birds. Individual human nerve fibers transmitting taste sensations respond differently to salty, bitter, sour, sweet, and savory—each sensing only one or two (and none responding directly to sweet stimuli)—and there is a great range of taste acuity and values among people. This information doubtless plays a role in taste tests on birds, in which some are found to have little or no response to bitter or sweet, while sour seems to be readily detected in small amounts, and food or water to which salt has been added may be preferred to the unaltered substance. Since most birds seem to have a limited sense of smell, and since the avian mouth and upper throat hold food only briefly during normal feeding in most birds (in marked contrast to humans), it is reasonable to assume that taste holds little of the subtle gratification for birds that it does for us. However, it may play a significant role in discriminating appropriate from inappropriate foods. It appears that birds feeding at winter roosts of Monarch butterflies in the central Mexican highlands may learn to detect the presence or absence of toxic cardiac glycosides by making quick "taste tests" on the insects' wings. This kind of discrimination is apparently rather limited, however, and no one has yet found a harmless substance that, for example, will make grain taste bad to flocks of blackbirds.

For how birds taste in another sense, see EDIBILITY. See also SMELL.

Torpidity

A slowing down or reduction in bodily function usually as a reaction to cold or stress or as a technique for conserving energy.

Torpidity in birds has been reported in three closely related families: the hummingbirds, the swifts, and the nightjars (and, in one instance, a species of swallow). In all cases, body temperature is greatly reduced, breathing and heartbeat become negligible, and the birds can be handled without arousing them.

Because of their small size and extremely high metabolism when active, most hummingbird species can barely sustain themselves overnight without feeding. To reduce energy requirements, some species of hummingbirds become dormant each night and return to normal at daylight, when they can resume feeding.

Flocks of White-throated Swifts of the western mountains of North America are known to roost in rock crevices in a torpid state during periods of cold weather when insect food is unavailable. This survival technique is also practiced by swift nestlings, which become torpid during the night in times of food scarcity.

The longest known period of torpidity—amounting to hibernation—is experienced by the Common Poorwill, which is known to "sleep" for nearly three months in rock crevices in the American Southwest. As with the swifts, the poorwill's torpidity coincides with cold weather and the consequent lack of insect food. Its temperature drops to 40°F (4.4°C) below normal, and it survives on stored fat. Though this is the only species of bird found hibernating in the wild so far, torpidity has been observed in other species of nightjars in captivity.

Treading
Refers to the male bird's action during copulation, based on the fact that in many cases, he stands on the back of the female. See SEX.

Twitching
British equivalent of the American term LISTING. In the UK, a "twitch" is a check-mark or "tick," and a twitcher is one whose chief interest in birds is checking

them off on personal lists (in British, "getting ticks"). According to birder/comic Bill Oddie, the term derives from such birders' emotional state over the prospect of seeing "a good bird": "He is so wracked with nervous anticipation (that he might see it) or trepidation (that he might miss it) that he literally twitches with the excitement of it all." According to Mark Cocker (*Birders: Tales of a Tribe*), the term arose on a specific occasion: an extremely avid twitcher rode his motorcycle through a night of sleet to pursue a vagrant Short-billed Dowitcher in England; he got his bird in the morning, but the icy journey resulted in some (temporary) physiological effects—he twitched uncontrollably.

See also BIRDWATCHING.

Vision (A bird's eye view)

The vertebrate eye is a very complex organ about which much remains to be learned. This entry is restricted to some conspicuous peculiarities of the avian eye and its function in comparison with the human eye.

THE AVIAN EYE IN GENERAL. Before offering generalities, it should first be noted that there is an enormous variation among eyes of different kinds of birds, so that the eye of a wren is as different from the eye of a goshawk as either is from the eye of an owl or a human. Nevertheless, it is broadly accurate to say that birds have proportionately *very large eyes and acute vision* and (with a few exceptions) depend on this sense more than any other. Though we normally see only a small portion of them, a bird's eyes take up far more room than any other part of its head; certain large predaceous birds have eyes as large as or larger than ours, and it has been stated that ostriches' eyes (diameter 2 inches; 5.1 cm) may be the largest of all terrestrial vertebrates. Another oft-quoted measure of avian eye size is that the eyes frequently outweigh the brain.

There is much disagreement about various types of visual competence (resolution, light sensitivity, focusing, etc.) in birds as compared with humans. There is no doubt, however, that some birds "see better" than we do—how much better is disputed—and that some have specialized powers, for example, highly developed night vision, good accommodation underwater, which we lack. It may also be true that many passerine species in a general sense see less well than we do. Some details are discussed below.

Birds require sight to find food, to orient themselves when traveling, and to perform basic functions, such as landing and flying between trees rather than into them. Other senses may assist these actions but cannot compensate for blindness. A bird that has lost both eyes is doomed, though there are cases of apparently healthy birds surviving with just one.

ABILITY TO DISTINGUISH DETAIL, technically called visual acuity or "resolving power," is best developed in birds (e.g., flycatchers and hawks) that must be able to see very small (or distant) moving objects. Birds with high acuity tend to have relatively large eyes with flat lenses at a greater distance from the retina—a combination that allows the projection of a larger image—and a very high concentration of specialized cells in the retina called cones, which allow acute perception of color and detail. It is widely published and repeated that the most keen-eyed day-flying raptors have resolving power eight times greater than that of a human, but this is now known to be an exaggeration. By comparing the number of cones in the retinas of raptors (about 1 million per sq. mm) with the human concentration (a mere 200,000 per sq. mm), some authorities deduce that the disparity is "only" 5×; but the most widely accepted estimate contends (on a different basis) that 2× or 3× is more like it. If you try to imagine seeing twice or three times "as well"—that is, acutely—as you now do, the visual competence of hawks and their relatives remains sufficiently impressive even by the most conservative standards. In fact, it is probably the most highly developed eyesight that exists.

The range of visual acuity in other types of birds ranges from significantly (perhaps 2× to 3×) poorer than ours to measurably—but not dramatically—better.

LIGHT-GATHERING ABILITY, or visual sensitivity, makes it possible to see well when very little light is present. It is most highly evolved in owls but is present to a significant degree (i.e., greater than in humans) in many birds. The eyes of owls are large and elongated (tubular) with comparatively wide corneal and lens surfaces, allowing a maximum amount of light to reach the retina. As would be expected, the concentration of light-sensitive "rods" is very high; cones are proportionately few, which means that owls have low visual acuity compared with day-flying raptors. (The "perceptor cells" [rods and cones] tend to be present in inverse proportions.)

Contrary to what is sometimes supposed, owls cannot see prey in total darkness, and at least some species depend heavily on sound as well as sight. It is also a mistake to believe that nocturnal owls are "blind," or nearly so, during the day simply because they are inactive; the visual acuity, at least of some species, is better than ours, night or day.

Night vision in birds, though in general superior to humans', usually takes longer to "turn on"—perhaps an hour or more as compared with about 10 minutes for us. Birds that make deep dives (e.g., loons) have increased rods for finding prey in nearly sunless depths.

FOCUSING, or "accommodation," is the ability to retain a sharp image at varying distances or under changing refractory conditions. It is achieved mainly by muscular action on the lens, stretching it flatter for seeing into the distance, making it more convex for examining the foreground. In birds, unlike mammals, the cornea also bends to aid focusing. A wide range of accommodation and the ability to focus rapidly are of special importance to fast-flying birds and particularly to those that make fast dives from the air to catch prey. Species such as cormorants, which make long dives underwater, must be able to focus through water as well as air. Birds with such special needs may have a focal range five times that of humans, but the capacity of more sedentary land birds may be less than or about equal to ours.

The chief factors that allow exceptional powers of accommodation in birds are (1) high development in the muscle apparatus (ciliary body), which controls the shape of the lens; (2) a very soft lens; and (3) a ring of small overlapping bones, the scleral ossicles (sclerotic ring), which stabilizes the eyeball while the lens is being pushed and squeezed.

Terns, which dive on fish from the air, cannot focus underwater and must rely on a "preset" aim. Penguins, by contrast, see well *only* in the water, where their eyes have evolved to follow fast-swimming fish.

SIZE OF FIELD AND DEPTH AND DISTANCE PERCEPTION. How much of the surrounding visible sphere one sees at a given time, the sense of three dimensionality, and judgment of relative distances depend on where the eyes are placed in the head, the extent to which they can be moved, and the ability (or lack of it) to turn the neck.

Human eyes are located close together in the front of the head. We therefore command a relatively narrow total field without moving the eyes or turning the neck. This field is significantly enlarged if we turn the neck without moving the rest of the body, but it is still by no means total. Because we receive a slightly different image in each eye, and the two images are combined by the brain (binocular vision), we readily perceive the depth as well as the length and width of objects (stereoscopic vision) and can tell how far away things are and the degree to which they get closer as we approach them. The field of binocular vision is, of course, always narrower than the total field, but the ratio of one to the other varies among different animals according to eye placement and other factors.

The eyes of owls are placed like humans' and have similar field (110° total; 70° binocular) and depth and distance perception. The fact that their eyes are rigidly fixed in their sockets is compensated for by an extraordinarily mobile neck. (This is true to a lesser extent of most birds.) The eyes of day-flying raptors are set somewhat more to the sides, so that their total field is wider (about 250°), but their binocular field is reduced to 35° to 50°. The eyes of most birds, however, are placed almost opposite each other on either side of the head. This gives them an enormous total field (up to 340°) but a very narrow binocular field (as low as 6° with an average of about 20° to 25°). Most birds, then, see most things with only one eye and have little depth perception. This may explain why ground-feeding birds will often cock their heads sideways—to see if that is really a seed or just a yellow spot on the ground—and why shorebirds (and others) turn one eye to the sky to appraise the threat posed by a form passing over, rather than cock their necks backward as we would do. It may also explain why some birds bob their heads when a potential threat approaches—to get a quick double-angle fix on an object in order to gauge its form and distance.

No discussion of visual field is complete without mentioning the woodcocks and the bitterns. The large eyes of the former are so set in the skull that, except for small "blind spots" directly in front and behind, their visual field is nearly total, even including the area directly above them; this may be an adaptation for perceiving danger from above by birds that spend much of their time focusing on the ground. The eyes of bitterns, on the other hand, are angled downward when the head is horizontal—so that they can see food on the ground without cocking or nodding the head—and forward when the head is upraised in their characteristic cryptic "freeze" posture. If you think this is easy, try looking someone in the face while your nose is pointed to the ceiling.

SEEING COLOR. Day-flying birds apparently have a significantly richer experience of color than we do. We deduce this from their sensitivity to the near ultraviolet spectrum, the presence of large numbers of cones containing visual pigments, and of colored oil droplets within the cones. These range from yellow to red and seem to act as filters, shutting out some of the blue/violet values and increasing sensitivity to yellow/red; it is interesting in this context to note the high percentage of avian display features that are yellow/red. The droplets may also help to cut glare and increase contrast, but there remains much to learn about their function.

Because color is perceived by the cones, nocturnal birds, which have relatively few, are believed to be color-blind.

The speculation that owls can perceive infrared light and thus "see" the heat generated by their prey in the dark is unproven.

IRIS COLOR AND EYESHINE. Birds, of course, have no eye whites as do humans, and eye color (exclusive of external eye rings) is that of the iris. In most birds the iris is dark brown or black, but a significant number of species have colored eyes. In some cases, this color is sexually dimorphic—males having the brighter color—and bright eye color is usually characteristic of adult birds and may heighten during the breeding season. The function of iris color is undetermined, but it may play a part in DISPLAY.

"Eyeshine" is a brilliant red or yellow (more rarely white or pale green) reflection emanating from the eyes of nightjars and other nocturnal birds and mammals when light hits them at a certain angle. This is produced by a thin iridescent membrane behind the retina called

Pauraque showing eyeshine

a tapetum, which shines through the colored but translucent surface of the retina.

THE NICTITATING MEMBRANE lies under the main eyelids of all birds and moves across the cornea at an angle from the lower inside of the eyeball (near the bill) to the upper outside. It is transparent in most birds (exceptions among nocturnal species) and used for regularly moistening the eye and perhaps for protection in the same way that we use our outer eyelids when we blink. (Birds rarely close their eyelids except to sleep.) The nictitating membrane of some diving birds is modified to aid accommodation underwater. It has been suggested that the transparent membrane may act as a "windshield" for birds in flight. Frogs, all reptiles, and some mammals (camels, seals, Polar Bear, and Aardvark) also have nictitating membranes.

Vocal Mimicry

It is well known that, in many cases, immature birds "perfect" their species' song by imitating their parents or other adult birds of their own kind and are also capable of improvising individual song variations. This kind of learning and ability reaches its greatest avian development in the so-called vocal mimics—birds that can (and habitually do), reproduce the sounds made by other birds and animals, including human speech, and even the "inanimate" sounds of the modern world.

Two types of vocal mimicry can be distinguished. The first consists of "natural" mimics, such as the New World mimic thrushes in the family Mimidae and the European Starling. Such species use imitated sounds along with motifs characteristic of their own species to create a unique individual song, the functions of which are analogous (with some differences; see Why They Do It, below) to normal territorial songs. This type of mimic may incorporate train whistles or other non-avian sounds, and can, if kept in captivity, imitate human speech, but normally has significant non-imitative components in its song as well. The composition and virtuosity of these songs vary widely among individuals and, to some extent, among different geographical populations.

The other types of avian mimics are the so-called talking birds. Certain members of the parrot family and the Common Hill Mynah (*Gracula religiosa*) are in such wide demand as pets because of their imitative talent (and, one suspects, the associated implication of empathy between owner and pet) that their wild populations are subject to significant pressure. Unlike the first type of mimic, talking birds never (as far as is known) use their mimetic faculty in the wild.

A plausible third category is represented by some jays that imitate the calls of hawks and crows but do not incorporate them into any kind of song. Some other corvids, for example, ravens, crows, and magpies, readily learn to "talk" in captivity.

Some natural mimics are much more versatile and enthusiastic imitators than others. The New World Northern Mockingbird may be in a class by itself, with a single bird known to have incorporated the songs of more than 80 other birds in its arias. And the number of sounds, including those of crickets, frogs, and cell phone ringtones that have been recorded for the species overall tops 400.

Among talking birds, the Grey Parrot (*Psittacus erithacus*) of Africa, the Yellow-headed Amazon (*Amazona ochrocephala*) of Middle America, and the little Budgerigar (*Melopsittacus undulatus)* of Australia are widely known as superior "conversationalists," while many other members of their family show little or no such ability. The Common Hill Mynah, which not only can repeat human words but also gives a very accurate rendering of a person's particular inflections and tone of voice, is probably supreme in this category. Many other mynahs and starlings and at least some crows exhibit some degree of imitative proficiency, though only a few (e.g., the European Starling) use this ability in the wild. It should be noted here that the cruel practice of splitting the tongues of captive birds to enhance their speaking ability does nothing to achieve this end and will soon result in the pet's death.

Northern Mockingbird and partial repertoire

HOW THEY DO IT. The best of the natural mimics have the greatest possible number of syringeal muscles (8–9 pairs), so that their physical ability to reproduce a wide variety of bird sounds does not seem very surprising. The imitation of human speech, however, is a different matter. Our ability to produce a wide range of vowel sounds is attributable to the ample resonating chambers of our throat, mouth, and nose; our tongue, teeth, and other mouth structures give us considerable versatility in producing consonants. Birds by contrast have few of these speech-producing modifications, and even our closest relatives, the great apes, which seem far better equipped than birds in these departments, never utter anything that resembles a word.

Some talking birds resolve the problem in part by "faking it," that is, using changes in pitch to simulate the variations in vowel and consonant sounds. This is abetted by human voice coaches, who can often hear their birds say things they have taught them with much greater clarity than an impartial auditor. These two factors in combination explain why "Budgie" says "Good morning, gorgeous," as clear as a bell to Aunt Lulu, while all you can hear is a jumble of psittacine squawks and gargles. Nevertheless, Hill Mynahs and other species can reproduce human speech with extraordinary fidelity, emitting vowels and consonants that are unmistakable even under audio spectrographic analysis. Exactly how they do this with their limited equipment is still poorly understood.

WHY THEY DO IT. Since many of the favorite talking birds never mimic any sounds in the wild and begin imitating speech only when confined in close contact with humans and deprived of association with their own species, it would seem that "talking" may be a form of social adjustment. Parrots are notably social and vocal birds and have been known to "pine" and die if deprived of association with their fellows. Therefore, talking may be a way of creating a necessary attachment to their captors. Once this is accepted, it is easy to understand apparently intelligent uses of language, such as speaking to people to whom they are attached, saying, "Come back here!" when they leave the room and making other apt or amusing comments—which are then reinforced by our appreciative responses.

The explanation for "natural" mimicry is not so apparent. Several thoughtful and plausible hypotheses have been advanced by ornithologists, but perhaps the best explanation of the mimetic impulse is that it is a form of *showing off* to a prospective mate or, to put it in the jargon of behavioral ecology, a "demonstration of fitness." As noted above, the most talented mockingbird males can imitate hundreds of sounds and do so mainly in the context of attracting and keeping mates—including in the midst of the reproductive act! And the largest repertoires seem to correlate with the greatest mating and reproductive success.

Finally, it should be mentioned that at least one appreciative listener has advanced the attractive theory that some mimics imitate sounds for "positive reinforcement" —that is, in essence, for fun.

See also SONG; INTELLIGENCE.

Wader

Wader
In most of the English-speaking birding world, a collective term for what American birdwatchers call "shorebirds"—which, it should be noted, excludes many birds of the shore (e.g., gulls and terns), while including a number of species which do not habitually visit the coast (e.g., woodcock and snipe). In North America, "wader" is a seldom-used term referring to long-legged waterbirds such as herons and ibises, also called "wading birds" in some texts.

Wetmore, Alexander (1886–1978)

Arguably the most versatile and prolific of all American ornithologists, Alexander Wetmore began modestly enough as a boy naturalist from North Freedom, Wisconsin. But he soon showed the promise of a distinguished future by publishing his first ornithological paper (on Red-headed Woodpecker behavior) at the age of 13. After earning a BA at the University of Kansas, he went to Washington, D.C., where he naturally fell in with Robert Ridgway and his circle of eminent zoologists at the National Museum. Wetmore took his PhD at Georgetown University and four years later (1924) was appointed Assistant Secretary of the Smithsonian Institution, head of the National Museum. Like other ornithologists of his stature, Wetmore presided over a number of professional organizations (president of the American Ornithologists' Union, 1926–1929) and received his share of awards. But the real measure of his career is his bibliography, which includes authoritative papers on avian physiology, pathology, behavior, distribution, migration, taxonomy, and paleontology (155 titles on the last subject alone). He also wrote the text for several popular bird books published by the National Geographic Society, but his *magnum opus* in book form is his four-volume *Birds of the Republic of Panama* (Smithsonian Misc. Coll. No. 150), for which he displayed his considerable skill as a field man and on which he was working at the time of his sudden death at the age of 92.

Wilson, Alexander (1766–1813)

Perhaps even more deserving than AUDUBON of the title "father of American ornithology," Wilson produced the first comprehensive, systematic, illustrated account of North American birdlife, the greatest natural history publication of its time in America. His *American Ornithology* in nine volumes was produced between 1808 and 1814, the last two volumes being published posthumously and the final one written by Wilson's friend, patron, and editor, George Ord. The work includes drawings of 320 individual birds in 76 plates, which amount, according to current taxonomy, to about 279 species.

Wilson's accomplishment is all the more impressive given his humble beginnings as the son of a weaver in Paisley, Scotland; the interruption of his education at the age of 13; his struggles to survive as a surveyor and teacher in Pennsylvania; and his complete ignorance of science and drawing prior to undertaking the writing and illustration of his *Ornithology*.

As with Audubon, some clue to his success may be found in his personality. He was a romantic poet in the tradition of Robert Burns and, as he said of himself in a letter to his future engraver, Alexander Lawson of Philadelphia, "long accustomed to the building of Airy Castles and brain Windmills." He was also possessed of strong convictions, which got him jailed in Scotland for writing political poetry in support of mill workers—probably largely responsible for his emigration from his homeland to America. His personal stubbornness and intolerance of criticism made him a touchy friend but were probably crucial to the completion of his life's work. He was also fortunate in attracting the support of well-placed patrons and advisors, such as William Bartram, prominent botanist and sometime ornithologist of the day; the famous painter and curator Charles Wilson Peale; Samuel E. Bradford, Wilson's employer as editor of *Ree's New Cyclopedia*, who eventually published *American Ornithology*; and George Ord, the wealthy and influential naturalist of Philadelphia, who championed Wilson's work, acted as his editor, wrote the final volume after Wilson's death, and defended his reputation against Audubon's inferences of ornithological plagiarism.

The famous rivalry between the two artist-naturalists arose (mainly after Wilson's death) from Audubon's assertion that Wilson's "Small-headed Flycatcher" was copied from his own drawing of 1808. Ironically, the controversy centered on an anonymous immature-warbler-like form which no one has seen (or at least recognized) since Audubon and/or Wilson encountered it. In defending his protégé's scruples, Ord pointed out with undeniable accuracy that a number of Audubon's portraits bear a striking likeness to some of Wilson's. Why Audubon, an

A Wilson's Warbler
eager to migrate

infinitely superior draftsman, should have been tempted to copy Wilson is mysterious, but the resemblance between his Bald Eagle, his Mississippi Kite, and his Red-winged Blackbirds and the indisputably prior drawings by Wilson of these species is beyond coincidence. Wilson died of dysentery and/or tuberculosis, age 47. His memory is amply preserved in ornithological nomenclature both scientific and vernacular, viz.: Wilson's

Storm-Petrel (*Oceanites oceanicus*); Wilson's Plover (*Charadrius wilsonia*); Wilson's Phalarope (*Phalaropus tricolor*); and Wilson's Warbler (*Cardellina pusilla*).

Zugunruhe

A German term meaning literally "travel restlessness"; applied by ornithologists to the fidgety anxiety exhibited by birds immediately before they begin to migrate. It is most conspicuous in migratory birds held captive during their normal migratory period.

Acknowledgments

No work of this kind can claim to be "original" in the sense that a novel or a scientific paper documenting new discoveries may be. It is essentially an amalgamation of other people's work, and the author's creativity is seen mainly in the selection of information, the organization of the material, and the words he chooses to summarize his sources. My main debt then is to the hundreds of ornithologists and other biologists and naturalists from whose work the *Birdpedia* is derived. By design, the *Birdpedia* contains few citations and no bibliography. Therefore, if the reader is eager to advance her quest for ornithological knowledge through deeper study, I (self-servingly) recommend the references and bibliography of my *Birdwatcher's Companion to North American Birdlife* (Princeton, 2004), the mother lode of information from which this slender volume is largely derived.

The above disclaimer to the contrary notwithstanding, there are a few individuals who deserve my explicit and profound gratitude:

Robert Kirk of Princeton University Press, who expertly edited the *Companion* cited above, invited me to undertake the *Birdpedia* and employed his many literary and diplomatic skills to shepherd the work to completion.

The late Edward S. (Ned) Brinkley, Editor of *North American Birds*, the American Birding Association's quarterly journal of ornithological record, provided invaluable advice in updating numerous matters of fact, taxonomy, and nomenclature as well as providing insights into the evolving customs and practices of the birding tribe. He also discreetly curbed not a few of my rhetorical excesses. His recent untimely death is a tragic loss to his many friends and colleagues and to the birding community as a whole.

Abby McBride has enlivened the text with 50 charming drawings, which combine accuracy with imagination—not as easy as you might think. The fact that we were forced by the COVID-19 virus to discuss her work solely by phone made the results especially impressive.

Finally, I must thank my immediate family, Kathy and Duncan Leahy, who had to endure my obsession with the book during almost five months of lockdown and acknowledge Luke Wolff Behringer for sharing his expertise on the owl fauna of Hogwarts School of Witchcraft and Wizardry.